SUPER SCIENCE with SIMPLE STUFF!

Activities for the Intermediate Grades

Susan Popelka

DALE
SEYMOUR
PUBLICATIONS®
P.O. BOX 10888
PALO ALTO, CA 94303

This book is published by Dale Seymour Publications®, an imprint of Addison Wesley Longman, Inc.

Managing Editor: Cathy Anderson
Project Editor: Patty Green Holubar
Production/Manufacturing Director: Janet Yearian
Production/Manufacturing Coordinator: Claire Flaherty
Design Manager: Jeff Kelly
Text and Cover Design: Remen-Willis Design Group
Text and Cover Illustrations: Pauline Phung

ISBN 0-201-49612-7
DS36837

4 5 6 7 8 9 10–ML–00 99

This Book Is Printed
On Recycled Paper

To my fantastic family—Carl, Erin, Mike, and Gail

Acknowledgments

There are many people whom I would like to thank for their assistance with this book. I would like to thank my family—Carl, Erin, Mike, and Gail Popelka—who helped and encouraged me in writing the book. The alliterations for the activity heads were sometimes elusive, but they made for great dinner table conversation. I would not have been able to finish the book without the daily question "Are you done with your book yet?" Thanks to a special four-year-old, who turned the computer over to me when I needed it and was very quiet while I was on the phone with the endless queries from the editor.

Speaking of the editor, thanks to Patty Holubar for her meticulous editing and wonderful suggestions for this book. I am grateful for the opportunity to have worked with someone on the same wavelength (check the Glossary for a definition); it made the editing of the book go very smoothly.

Thanks to Jeff Rosenthal, the science editor, for his thoughtful and careful check of the science explanations in the manuscript.

For the past 10 years, I have been teaching hands-on science workshops to elementary teachers. This book is the direct result of those teachers asking me to write it. The activities in the book are their favorite ones, and they have tried and tested all of them. I owe them a heartfelt thank-you for teaching me as I was teaching them.

Thanks to all my past teachers, especially my science and writing teachers. The skills they taught me have enabled me to pursue a very enjoyable and rewarding career.

CONTENTS

INTRODUCTION

How to Learn Science

The basic premise of this book comes from what a high school science teacher once told me: "Science is less scary when it is done with less-scary stuff." If you can do science activities using "simple stuff" that is found around the house and that you feel comfortable with, then science will seem interesting, fun, and understandable. The activities in this book use simple stuff that is readily available at home or can be purchased, for example, at a grocery store, hardware store, or school supply store.

The recent trend in science education has been to introduce more hands-on activities, like the ones in this book, into the classroom. Hands-on activities allow students to experiment and to manipulate materials, and they foster teamwork and cooperative skills as students work together doing the activities. Ultimately, hands-on activities lead to lively discussions about the scientific principles involved and to more experimentation as students ask "What will happen if . . . ?"

Hands-on science activities, however, must be presented in the context not only of having fun by manipulating materials but of understanding what is going on conceptually. The activities need to be accompanied by a well-thought-out plan for making predictions, gathering and recording data, and drawing conclusions. In this book, even the activity of making ice cream is accompanied by a student page asking the students to make and record observations and to explain "what it means."

A Chinese proverb that is often quoted to summarize the point of hands-on science activities is

> I hear, and I forget.
>
> I see, and I remember.
>
> I do, and I understand.

To this proverb, I would add a final line: "I explain what it means, and I *really* understand."

How This Book Is Set Up

This book is not designed as a stand-alone science curriculum. It is intended to be used as a resource book to augment your existing science curriculum. You probably won't use all the activities in the book but only those that apply to the science concepts you teach.

The activities in this book are designed for students in grades three through six. Many of the activities, however, can be adapted for lower grades or higher grades. The vocabulary you use and the students' level of understanding will reflect the grade level at which the activity is presented.

The order of the chapters in the book is based on the following reasoning: Motion is basic to the understanding of all physical science, so it is presented first—just as it is in most physics texts. Heat is energy that results from motion, so the heat chapter naturally comes after the motion chapter. Electricity and magnetism are presented back-to-back because so many of the concepts are interrelated. Likewise, many of the same wave concepts are used in discussing sound and light, so those chapters are also presented back-to-back. "Why do airplanes fly?" and "Why do boats float?" are two of the most-common physical science questions asked by children (and by adults, for that matter). The concepts involved in answering those questions are presented in back-to-back chapters on air pressure and buoyancy. The activities in the last two chapters—on center of gravity and chemical reactions—focus on concepts that could actually be included in other chapters in the book. They were, however, grouped in the chapters you see because of a common thread.

Within each chapter, the order of activities is based on the concept, with activities presenting similar concepts being grouped together. But even though the chapters are grouped logically and the activities are grouped by concept within chapters, the activities can be used as stand-alone activities. You can pick any activity in the book and do it successfully without having done any other particular activity before it. You can start any chapter without doing the chapters before it, too.

You will notice that some of the activities could be placed in more than one chapter by virtue of the fact that they involve several concepts. Some heat activities could be grouped with air pressure, for example. When you present activities that involve several concepts, it is a good idea to follow the student page, which focuses on the key concept(s) being taught in the activity. After that, you may want to point out that there is more science going on and then challenge the students to figure out what other concepts are involved. If the activity presents concepts you have already discussed in your science class, take the opportunity to review those concepts in light of the new activity. If the activity presents concepts that will be taught later, it's a good idea to wait until that time and use the activity to introduce the new concepts.

There are certain key science concepts that appear over and over again in different chapters. Frequency is an example of such a concept; it is related to light, sound, and motion. Recognizing and understanding that some concepts are woven through many activities will help your students better grasp those scientific principles and perceive the interconnectedness of all science.

Overview

Each chapter opens with an overview of the key concepts presented in the chapter's activities. You should read through this overview before doing an activity from the chapter, in order to get general background information about the chapter. Many of the technical terms used in the chapter are described in the overview. You may find it useful to refer to the Glossary for additional information.

Teacher page

For each activity in the book, there are a teacher page and a student page. The teacher page provides you with background information and

directions for performing the activity. The scientific concept(s) focused on in the activity is given in the Science section. This section guides you in picking which activities will fit into your science curriculum and where they will fit. The concept(s) is stated briefly and simply; the explanation of the science comes later in What's Going On Here.

Stuff lists materials needed to do the activity. Most of the materials are available at home or can be purchased at a grocery, hardware, or school supply store. In the few cases in which materials are not readily available, suggestions about obtaining them are given. You will notice that the Stuff list uses measurements that are in the English system rather than the more-scientific metric system. Most science books use the metric system, but most grocery stores and hardware stores in the United States don't. To make it easier for you to purchase and prepare the stuff you need for the activities, the measurements are given in the most-convenient (albeit not scientific) terms. When you are doing the activity with your students, you may choose to use the metric system in order to be consistent with your science curriculum. Or you may choose to use the English system, reminding your students that there are many ways to measure things.

Step-by-step instructions for the activity are given in the What to Do section. The *you* in the instructions refers to you the teacher. The instructions assume that you are reading the page while doing the activity before presenting it in the classroom. When presenting the activity in the classroom, you may do either of the following:

- Give the directions verbally, reading from the teacher page.
- Summarize the directions in your own words as the students do the activity.

A brief explanation of the main scientific concept(s) comes next in What's Going On Here. Information in this section will help you guide your students as they answer questions on the student page. You may run across some technical terms in this section that you will want to look up in the Glossary.

A number of ideas for extending the activity are suggested in Try It! Sometimes the suggested changes in the activity will yield similar results; sometimes the results will be better; and sometimes you won't get good results at all. The point of this section is to encourage the students to try new things and to determine what effect the new stuff used or the new procedure followed has on the activity. As in any good science activity, you must remember to change only one thing (for example, material, temperature, or quantity) at a time.

Group-size icons, described below, appear on the teacher page for each activity to recommend the optimum group size for the activity; the recommendations range from teacher-directed demonstrations to small groups and individual work. You, however, know your classroom resources and your students best; the group size that you use will ultimately depend on the availability and cost of materials and also on the developmental skills of your students.

Note that the "stuff" listed for each activity consists of the materials needed for *one* group or individual. As you gather supplies for an activity, you will need to multiply the materials by the number of groups or individuals who will be doing the activity.

 Teacher-directed activities These activities should be done by you as demonstrations. Most of the activities that are teacher directed have some safety issue associated with them or require materials that may be impractical to obtain for small groups or individuals. These activities have been chosen and designed to be appropriate for demonstrating in front of a group of students. However, even in teacher demonstrations, try to get the students involved by having them assist with the activity. This helps demystify science (teachers aren't the only ones who can do it), and the students also gain confidence and experience in presenting science to their classmates.

 Small-group activities These activities should be done in small groups; the actual size of each group will depend on the class size and the availability of the needed materials. When the students work on science activities in small groups, they should exchange roles often so that everyone has the opportunity to manipulate the hands-on materials. Have the students do the activities in small groups or individually whenever possible (keep in mind the Chinese proverb referred to at the beginning of this Introduction).

 Individual activities These activities are best done individually. Often the activity results in something that the students are able to take home. For some individual activities, the observations that are made would be hard to notice in a group activity or teacher-directed demonstration. Even so, there is much that can be learned from working together in groups, and the discussions that take place as students work on science activities together is itself a learning experience. Thus, even if the activities are done individually, it is still a good idea to have the students work together in small groups doing them.

Also on the teacher page for each activity appear safety icons, described below. The activities in this book are designed for elementary students. They should be done with adult supervision, using the materials that are recommended and carefully following directions. However, as a teacher, you know your class best and you know what they are capable of doing. You may need to adapt some of the activities to suit your classroom: For example, you may need to prepare ahead of time materials that require using pointed scissors. Or you may decide that your students can prepare all the materials that are needed. Whatever your decision, you and the students should pay attention to the following safety precautions:

 Sharp objects In these activities, the pointed end of a pair of scissors or a pencil may be used to make a hole in something. Knives and needles are also used to prepare some materials. When using any of these sharp objects, care should be taken to avoid injury. Watch out for soup cans that have rough edges that can cut skin. Glass plates and mirrors also often have rough edges that can cut skin; tape the edges ahead of time for the students, and be careful when handling the glass or mirrors. Wear goggles when working with sharp objects.

Heat and fire Use caution near open flames, and wear goggles. When using hot tap water, make sure that the water is below 104°F. Use insulated gloves when working with boiling water and when using batteries that get hot.

 Flying objects Objects that are twirled need open space away from people. Objects should be able to be twirled without hitting anything. Thrown or launched objects need an area that is free from obstructions; care should be taken so that no one is in the way when an object is thrown or launched. Goggles should be worn.

 Chemicals All the chemicals used in the activities in this book are available at the grocery store. But caution must still be used when working with them. Read the warning on the containers of any chemicals that you use. Wear goggles to protect your eyes. Never taste chemicals. Avoid splashing chemicals on your skin or clothing. Wash your hands thoroughly with soap and water after using any chemicals.

 Magnets Keep magnets away from computers, computer disks, credit cards, videotapes, audiotapes, televisions, video recorders, tape recorders, telephones, answering machines, radios, and loudspeakers.

 Noise Do not speak loudly or shout into someone's ear or into anything that is attached to someone's ear. Do not blow into someone's ear or into anything that is attached to someone's ear.

Student page

The student page is written for the students to read and work on as you guide the activity. The *you* on the student page refers to the student.

You will discover that the student page follows the scientific method. The What You Want to Know section is a *statement of the problem* in the form of one or more questions. What You Think Will Happen allows the students to make predictions, or *hypothesize*, usually by completing one or more multiple-choice statements. The What Happened section requires the students to

gather and record data and/or make observations. What It Means is the *conclusion* that the students draw on the basis of their observations.

As the students fill out the page, they often will be writing the same information in different sections. This redundancy is built into the page to reinforce the scientific concepts involved in the hands-on science activities.

Most of the student pages have a drawing that describes the activity. The drawing serves as a visual reminder of the activity, and it also is handy when the students take the page home and explain the activity to others. A few student pages do not have drawings on them because even a simple drawing would take away the feeling of discovery during the activity. In these cases, the students should be encouraged to draw their own pictures of the activity on the back of the student page after doing the activity.

What You Want to Know is one or more questions that give the students an idea of what the activity is all about. It suggests a purpose for doing the activity and starts the students wondering about what they will learn.

Prediction is an important skill to be learned in any science program. Most of the activities in this book have multiple-choice statements that require the students to predict "what you think will happen." Multiple-choice statements are used instead of open-ended questions so that the students are able to focus their thoughts and so that they gain confidence in making predictions. The students should be told that there are no right and wrong answers when it comes to making predictions. The multiple-choice statements are written so that the students can pick more than one answer; in fact, you should encourage the students to pick as many answers as they think appropriate. Some multiple-choice statements even have a blank answer line, allowing the students to write in their own prediction. There may appear to be clues to the answers to these questions in the title of the activity, in the drawings, or in some of the questions that follow. The students should be encouraged to look for such clues to make their predictions; that is an important skill in science.

In What Happened, the students make observations and record their data. Activities have questions to answer or tables to fill in with the observations from the activity. In this section, the students should be encouraged to record exactly what they observed; they should not draw any conclusions at this point or try to make their observation fit what they thought would happen.

What It Means usually starts out with the phrase "what do your observations tell you about" or "what can you now say about." This section requires the students to draw conclusions based on what they observed in the activity; they make deductions using their own senses. The aim here is to relate what happened in the activity to the scientific concept that is involved. You should use information from the teacher page Science and What's Going On Here sections to help guide the students. Depending on the experience and background of your students, you may have to guide them step by step through some or all the sections on the student page.

Assessment

The activities in this book are designed to supplement an existing science curriculum in which many types of assessment are used. Standardized paper-and-pencil tests are not appropriate for hands-on activities. But that does not mean that assessment cannot or should not be used. In fact, embedded assessment is designed into every activity in this book. The questions in the What It Means section of the student page have been formulated so that the students explain the meaning of what they observed; their statements about the key concept(s) of each activity provide the basis for embedded assessment. Students should not be assessed by the accuracy of the results they get, but rather by the process by which they get their results. As the students are doing the activity, you may also assess their skills in working in groups and in using the scientific method.

Glossary

The Glossary contains easy-to-understand definitions for the terms used in the What's Going On Here section on the teacher page. Often the definition of a term will also include an example to

help explain it. The Glossary is written for the teacher, but most of the definitions can be used as is or with slight modifications for middle elementary students.

You will find that sometimes a definition for a term is given on the student page. For example, the definition for *density* appears on several student pages. Definitions used on the student page are reserved for those terms that seem to be most often misunderstood by students.

Selected bibliography

The Selected Bibliography is a list of science activity books appropriate for the middle elementary grades. The books in the list will give you ideas for more scientific activities to use in your classroom. If you are doing a unit on magnets, for example, and run out of activities from this book, check out a book from the bibliography. Some of the books in the bibliography contain activities that are similar to the activities in this book but take a different approach. If your students become really interested in a particular activity, you may want to try to find a similar one in one of these books and let them try the activity from a different perspective.

Index

The Index consists mainly of entries for the concepts covered in the book. If you are teaching a unit on light, you will discover that it is easy to find the activities dealing with various aspects of light by simply referring to the entry "light." If you want to locate an activity dealing specifically with reflection of light, you will be able to find those activities by looking under "reflection" in the Index. A few entries in the Index are activity related—for example, "ice cream." The activity-related entries have been included to make it easy for you to find the few activities that are best remembered not by their topic area (temperature, in the case of ice cream) but by the final (delicious, in the case of ice cream) product.

How to Use the Book

There are many ways to do the activities in this book. Your individual teaching style will determine what method works best for you. The class size and classroom environment will also determine how the activities are presented. It is a good idea to allow the students to do as much of an activity as they are capable of doing. The results that they get may not look as refined as what adults can do, but the sense of ownership and accomplishment will far outweigh any lack of finesse. What follows is a suggested presentation format:

Before presenting the activity in the classroom,

1. Try the activity at home or school. Find a place to do the activities that can get messy. Gather the materials; follow the directions carefully; and observe what happens. Take notes on the activity.

2. Make predictions as you proceed from one step to the next. Ask yourself the questions "What will happen if . . . ?" or "What will happen when . . . ?"

3. Try a few of the ideas in the Try It! section.

4. Decide how you will introduce the activity in the classroom. Review the student page so that you understand how the science concepts are integrated into the activity.

5. Most of the activities in this book take between 30 minutes and an hour, but your individual teaching style and other factors will affect the actual time that any activity takes. Decide how much time you will need to do the activity in your classroom.

6. Decide what group size—whole class, small groups, or individuals—will be used when doing the activity.

7. Gather the materials that will be needed (one set of materials for each individual or group). Plan to have extras of items that may break or roll under cabinets. Bring any other items that the students may need if they want to attempt the ideas in the Try It! section.

8. Make the required number of copies of the student page.

When presenting the activity in the classroom,

1. Hand out the student pages. If appropriate, divide the students into groups.

2. Have the students read the question(s) in the What You Want to Know section.

3. Describe the activity to the students, and have them make a prediction in the What You Think Will Happen section. Discuss the predictions. Do not be critical of any predictions; in order for students to take the risk that is sometimes necessary to make predictions, they need to feel comfortable that they will sometimes be wrong and that that is all right. What is more important than making a correct prediction is that they use their own background and previous experience to make an educated guess (another term for *prediction* or *hypothesis*). Discourage the students from haphazardly picking any choice simply to get on with the activity. Encourage them to explain why they made the prediction that they did. Different predictions from various students will facilitate good discussions.

4. If the students are working individually or in small groups, hand out or have the students pick up the materials. In cooperative groups, it often works well to have one person gather the materials.

5. Read or summarize the directions for the students as you guide them through the activity or do it in front of them.

6. Have the students record the required information from the activity on their pages in the What Happened section. In order for the scientific concepts to really sink in, it is important that individual students fill out their own page. (In some activities, the students record data in a table with two columns of numbers. When that happens, it is always a good idea to graph the data. Graphs give meaning to tables of numbers. After the students graph the results from the activity, discuss the graph and have the students write a few sentences describing what they learned from it.)

7. Discuss the activity with the whole class before the students fill out the What It Means section. Reading the Science and What's Going On Here sections from the teacher page will help you generate discussion questions and help you guide the students to an understanding of how what they observed relates to answering the What It Means question(s). Encourage the students to use the blank side of their student page to write on if they do not have enough room on the front. If appropriate to the activity, you may also want to encourage the students to use the blank side of the student page to draw a picture of the activity.

8. After the students have completed the activity, ask them what else they could try to expand the activity or to answer questions that came up while they were doing the activity. Ask the students to complete the question "What would happen if . . . ?" Provide them the opportunity to continue their investigation immediately if time and materials allow; if that's not feasible, suggest a plan to continue the activity the following day.

9. Ask the students if they would be interested in trying any of the suggested activities in the Try It! section on the teacher page. To motivate the students, you can preface each suggestion with "I wonder what would happen if we did this activity again and changed . . . " If you have the materials available in the classroom and time permits, try the activity again right away, making the necessary changes. Otherwise, you can try it in class the next day. Some of the Try It! activities, incidentally, make excellent science fair projects.

All the activities in this book have been pretested by teachers, elementary students, and the author. You may, however, encounter an activity that does not work the first time. If that happens to you, relax and do the following:

- Make sure that you have all the recommended stuff.
- Try the activity again, following the instructions very carefully.
- Celebrate; you may have discovered a new scientific principle!

Hands-on science can be messy, and it can also be unpredictable. That is why it is so much fun! Sometimes an activity that worked perfectly well on the kitchen table at home will totally fail in the classroom with a group of students. So if things don't always work out as you anticipated, console yourself in the knowledge that the basic science principles always govern how things work. Science never fails when an activity does; it is the materials or procedures that are inadequate. (Rarely does the scientist goof!)

I hope that you enjoy doing the activities in this book, learn more science, and have a lot of fun doing it. I certainly have!

CHAPTER 1 ■ MOTION

Memorable Motion Matters

- In this chapter, you will explore some interesting concepts about motion that were discovered by two famous scientists. Sir Isaac Newton (1642–1727) was an English astronomer, scientist, and mathematician who made many important discoveries about motion and light. Galileo (1564–1642) was an Italian astronomer and scientist who studied falling bodies and discovered the law of the pendulum.

- Newton's first law of motion (also called the law of inertia): An object moves in a straight line at a constant speed unless some force (a push or a pull) comes along and changes its direction or speed. An object will also stay still unless something comes along and moves it.

- A consequence of Newton's first law is that a force is needed to make an object move in a circle because the direction that the object is moving is constantly changing.

- Newton's second law of motion: The more you push or pull on an object, the greater effect you have on changing its direction or speed of motion. Also, the object will move in the same direction that you pushed or pulled it.

- Newton's third law of motion: For every force on one object, there is an equal (in size) and opposite (in direction) force on another object.

- The law of gravity (also courtesy of Sir Isaac Newton): Every object in the universe attracts every other object in the universe with a gravitational force. This force is greater for greater masses, and it decreases as the distance between the objects increases.

- You are familiar with the force of gravity acting on you by the earth. The earth is pulling you downward, toward its center, with a force that is equal to your weight. But by Newton's third law, you are also pulling the earth upward with the same force.
- If two objects of unequal weight are dropped from the same height at the same time, they will hit the ground at the same time—if gravity is the only force acting on them.
- There are a couple of forces—friction and air resistance—that oppose the motion of objects.
- The force of friction is caused by irregularities in the surface of objects that are moving over each other. The direction of the force of friction is always opposite to the motion of an object.
- The molecules of air keep objects from moving through it freely; as an object moves through air, it is constantly bumping into air molecules that slow it down. This is called the force of air resistance, and it depends on the shape, size, and speed of the object. Air resistance is what causes a feather to float slowly to the ground and a penny to drop very quickly.
- Centripetal force is the force on an object that is moving in a circular path. The direction of this force is toward the center of the circle.

FREE FALL

SCIENCE: Air resistance and gravity affect the motion of a falling object.

STUFF: Cotton ball, marble, two small plastic containers with lids (for example, 35-millimeter-film canisters)

What to Do

1. With your hands above your head, drop a cotton ball and a marble from the same height at the same time. Observe which one hits the floor first.
2. Place the cotton ball in one container, and put the lid on it. Place the marble in the other container, and put the lid on it.

3. Again with your hands above your head, drop the two containers from the same height at the same time. Observe which one hits the floor first.

What's Going On Here

In the early 1600s, Galileo stated that the rate at which an object falls toward the earth does not depend on its mass. The cotton ball and the marble should fall together at the same rate and hit the floor at the same time; this is not what you observe because there is another force at work here besides gravity. That force is called air resistance, and it depends on the shape of the object and the speed at which the object is moving. Objects that are spread out, like the cotton ball, experience more air resistance than do those that are packed together, like the marble. Thus the marble hits the floor before the cotton ball. When you place the cotton ball and the marble in individual containers and drop them, they fall at the same rate. The force of air resistance on both containers is essentially the same, and they fall together, hitting the floor at the same time.

Try It!

Repeat the experiment on the moon. Just kidding! Astronauts already did the experiment on the moon and discovered that a hammer and a feather dropped from the same height at the same time fell at the same rate and hit the ground together. There is virtually no atmosphere on the moon and therefore no air resistance.

Measure the height that the cotton ball is dropped from and the time it takes to hit the floor. Divide the height by the time to get its average speed.

Drop a crumpled-up piece of paper and a flat piece of paper, and see which one hits the floor first.

Try dropping other objects.

FREE FALL

What You Want to Know

If two objects of different weights are dropped from the same height at the same time, which hits the floor first?

What You Think Will Happen

If you drop a cotton ball and a marble from the same height at the same time,

        a. the marble will hit the floor first.

        b. the cotton ball will hit the floor first.

        c. both objects will hit at the same time.

If you place the cotton ball in a container with a lid and the marble in a similar container with a lid and drop them again from the same height at the same time,

        a. the container with the marble will hit the floor first.

        b. the container with the cotton ball will hit the floor first.

        c. both containers will hit at the same time.

What Happened

Which landed first when you dropped the cotton ball and the marble from the same height at the same time?

_____

_____

Which landed first when you dropped the containers with the cotton ball and the marble from the same height at the same time?

_____

What It Means

What do your observations tell you about what happens when objects of different weights fall toward the floor?

_____

What could you do to one of the containers to make it fall more slowly than the other one?

_____

PLUNGING PARACHUTES

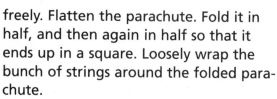

SCIENCE: Air resistance slows down a falling object.

STUFF: Clothespin, stopwatch (optional), string (160 inches long), ruler, scissors, 6-inch and 12-inch squares of the same fabric

What to Do

1. Throw a clothespin up in the air a few feet. Use a stopwatch, or count the seconds it takes to fall to the floor.
2. Cut four strings, each about 20 inches long. Tie each string to a corner of a 6-inch square of fabric.
3. Tie the four free ends of the strings together in a knot. Be sure the strings are all the same length as measured from the fabric square.
4. Tie the knotted end of the strings to the top of the clothespin.
5. Grasp the center of the parachute material, allowing the clothespin to hang freely. Flatten the parachute. Fold it in half, and then again in half so that it ends up in a square. Loosely wrap the bunch of strings around the folded parachute.
6. Throw the parachute up in the air. Try to throw it the same height as in Step 1. Use a stopwatch, or count the seconds the parachute takes to fall to the floor.
7. Repeat Steps 2–6, using a 12-inch square of fabric. Try to throw the parachute the same height as in Step 1.

What's Going On Here

A falling object has two forces acting on it. The force of gravity pulls the object downward toward the earth. As the object falls downward, the air pushes upward on it with a force called air resistance. If the weight of an object is greater than the force of air resistance, the object speeds up as it falls, going faster and faster. Objects with a large surface, such as the parachute, experience more air resistance than objects with less surface area. If the object has a large surface and small weight, the upward push of the air can equal the downward force due to weight, causing the object to float gently downward like a feather.

Try It!

Try different materials for the parachute (for example, tissue paper or garbage bags).

Try different lengths of string.

Try cutting small holes in the parachute.

Try using the parachute on different objects.

Try a different size of parachute material.

Try a different shape for the parachute.

PLUNGING PARACHUTES

What You Want to Know
How does the time a clothespin takes to fall to the floor change when you attach a homemade parachute to it? How does the size of the parachute affect the time it takes to fall to the floor?

What You Think Will Happen
When you attach a parachute to a clothespin, the clothespin will
   a. take longer to fall to the floor.
   b. fall more quickly to the floor.
   c. fall to the floor in about the same time as when it did not have a parachute.

With a larger parachute, the clothespin will
   a. take longer to fall to the floor than with a smaller parachute.
   b. fall to the floor more quickly than with a smaller parachute.
   c. take about the same time to fall to the floor as with a smaller parachute.

What Happened
Record your observations in the table.

Clothespin	Time to fall to floor
With no parachute	
With 6-inch parachute	
With 12-inch parachute	

What It Means
What can you now say about how a parachute affects the time it takes a clothespin to fall to the floor?

_____

What can you now say about how the size of a parachute affects the time it takes a clothespin to fall to the floor?

_____

STAY STANDING STILL

SCIENCE: An object at rest tends to stay at rest.

STUFF: Strip of notebook paper (1 inch x 11½ inches), marker that will stand by itself

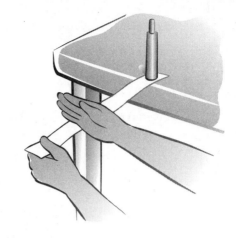

What to Do

1. Place a strip of paper so that about an inch lies on a table and the rest hangs off the edge of the table.
2. Stand a marker on the end of the strip of paper on the table.
3. Grasp the strip of paper by its free end, and *very slowly* pull the strip of paper away from the table. The marker will fall.

4. Repeat Steps 1 and 2.
5. This time, with one hand grasp the strip of paper by its free end, and with your other hand quickly strike a point on the paper midway between the marker and the end of the paper you are grabbing. The marker should stay standing still.

What's Going On Here

There is friction between the paper and the marker. When you pull the paper very slowly, friction keeps the paper in contact with the marker, and the marker moves with the paper until it falls over. When you strike the paper very quickly, friction between the paper and the marker is over-come, and the marker remains standing as the paper is pulled away. Part of Newton's first law of motion says that an object at rest will remain at rest. The marker is at rest, and even though the paper is moved, the marker remains standing still.

 **Try It!**

Try using different things for the standing-still object (for example, marble, building block, soft drink can).

Try using strips of different things for the standing-still object to stand on (for example, sandpaper, wax paper, aluminum foil, fabric).

Try using a strip of paper about 3 inches by 11 inches and an unopened soft drink can as the standing-still object. Then try the activity with an empty soft drink can. Try to determine which can requires less force striking the paper for the can to remain standing still.

STAY STANDING STILL

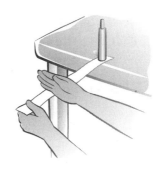

What You Want to Know
What happens when you slowly pull a strip of paper out from under a marker that is standing on the end of the paper? What happens when you rapidly pull the strip of paper?

What You Think Will Happen
When you *slowly* pull a strip of paper out from under a marker that is standing on the end of the paper,

   a. the marker will fall slowly.
   b. the marker will fall quickly.
   c. the marker will remain standing.
   d. _____.

When you *rapidly* pull a strip of paper out from under a marker that is standing on the end of the paper,

   a. the marker will fall slowly.
   b. the marker will fall quickly.
   c. the marker will remain standing.
   d. _____.

What Happened
What happened when you *slowly* pulled the strip of paper out from under the marker standing on the end of the paper?

_____

What happened when you *rapidly* pulled the strip of paper out from under the marker that was standing on the end of the paper?

_____

What It Means
A magic trick is pulling a tablecloth from under dishes on a table without the dishes falling on the floor. What did you learn in this activity that could help you do that trick?

_____

_____

ROLLICKING ROLLER

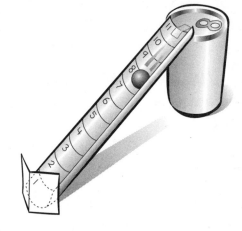

SCIENCE: The energy of a rolling object depends on its speed.

STUFF: Paper towel tube, scissors, ruler, marker, several books or a soft drink can, masking tape, two index cards (3 inches x 5 inches), marble

What to Do

1. Cut a paper towel tube in half length-wise to form a trough.
2. Make a mark on the trough every inch. Label the marks "1" for 1 inch, "2" for 2 inches, and so on.
3. Work on a flat surface such as a table. Prop the trough up on books or a soft drink can so that the top of the trough is about 4 or 5 inches above the table. The end of the trough near the 1-inch mark should touch the table. Tape the trough to the books or soft drink can and to the table so that the trough is clear and is securely fastened.
4. Holding two index cards together, fold them in half the short way and tape

them together. Open the taped cards halfway to make a target that resembles a greeting card.
5. Place the target touching the end of the trough so that when a marble is rolled down the trough, it will hit the inside wedge of the target and move it some distance.
6. Place the marble at the 1-inch mark on the trough and allow it to roll down the trough.
7. Measure the distance that the target moved from the end of the trough.
8. Repeat Steps 5–7 for the other distances marked on the trough.

What's Going On Here

As the marble rolls down the trough, the force of gravity increases its speed. Because of its speed, the marble has kinetic energy and is able to do the work of moving the target at the bottom of the trough. The greater the height the marble is rolled from, the more speed it has and thus the more energy it has to move the target.

Try It!

Try using a longer tube (for example, a wrapping-paper tube).

Try propping the tube up higher (or lower) so that it has more (or less) slope.

Try a smaller or a larger target.

Try placing the target farther from the end of the tube.

ROLLICKING ROLLER

What You Want to Know

When a marble is rolled down a slope, how does the height that it is rolled from affect how far it can move a target at the bottom of the slope?

What You Think Will Happen

You roll a marble down a slope, and it strikes a target at the bottom of the slope, moving the target some distance. If you roll the marble from a higher point, the target will move

       a. the same distance as when you roll from a lower height.

       b. a shorter distance than when you roll from a lower point.

       c. a longer distance than when you roll from a lower point.

What Happened

Record your observations in the table.

Trough mark for marble release	Distance target moved
1 inch	
2 inches	
3 inches	
4 inches	
5 inches	
6 inches	
7 inches	
8 inches	
9 inches	
10 inches	

What It Means

What do your observations tell you about how the height from which a marble is rolled down a slope affects the distance that it can move a target at the bottom of the slope?

_____

ROLLICKING ROLLER, REVISITED

SCIENCE: An object can travel farther on a smooth surface than on a rough one because the rough surface has more friction.

STUFF: Paper towel tube, scissors, book, ruler, masking tape, six samples of smooth and rough surfaces (6 inches x 12 inches; for example, aluminum foil, carpet, sandpaper, wax paper, several kinds of fabric), marble

What to Do

1. Cut a paper towel tube in half lengthwise to form a trough.
2. Work on a flat surface such as a table. Prop the trough up on a book so that the top of the trough is about 1 inch above the table. Tape the trough to the book, and rest the other end on a piece of aluminum foil.
3. Place a marble about halfway up the trough, and allow it to roll down the trough without giving it a push.

4. Measure the distance from the end of the trough that the marble travels before it stops. If the marble rolls too far, you may have to change the height of the trough or the point in the trough from which the marble starts rolling.
5. Change the sample at the end of the trough, and repeat Steps 3 and 4. Use the same height for the trough and point on the trough from which the marble rolls that you settled on in Step 4. Do this for each sample.

What's Going On Here

Friction is a force that opposes the motion of an object. The rolling marble has friction acting on it during its journey down the trough as well as on the sample. The friction on it in the trough is the same from trial to trial, but the friction encountered after leaving the trough changes depending on the sample at the end of the trough. The marble will travel farther on smooth surfaces because they offer less friction than rough surfaces.

Try It!

Try lining the trough with different materials and measuring how far the marble travels on a smooth table after leaving the trough.

Try using different rolling objects like golf balls, table tennis balls, or rolled balls of aluminum foil. You may have to change the size of the trough.

Try placing different samples end to end so that the marble can travel from one to another before it stops. Will it speed up when it goes from a rough surface to a smooth one?

ROLLICKING ROLLER, REVISITED

What You Want to Know
Does a marble travel farther on a smooth surface or on a rough surface?

What You Think Will Happen
A marble is rolled down a slope and hits a surface at the end of the slope on which it can travel horizontally. The marble will travel

        a. farther on a rough surface.
        b. farther on a smooth surface.
        c. the same distance on both surfaces.

What Happened
Record your observations in the table.

Surface	Distance marble traveled

What It Means
What can you now say about the type of surface on which a marble can travel farthest?

_____

What do you think causes the marble to move differently on different surfaces?

_____

What could you do to the trough to make the marble travel farther after leaving it?

_____

_____

SWIFT SWINGER

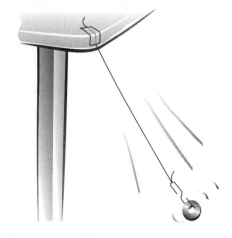

SCIENCE: The time it takes for a pendulum to swing back and forth depends on the length of the pendulum but not on its mass.

STUFF: Paper clip, string (24 inches long), tape, 10 washers, stopwatch, ruler

What to Do

1. Tie a paper clip to one end of a string. Tape the other end of the string to a corner of a table so that the paper clip can swing freely when it is pulled back.
2. Open the paper clip up so that washers can be slid onto it. Place one washer on the paper clip. Pull the washer back a few inches and let it go. Time five complete swings of this pendulum. (A *complete swing* is the motion of the pendulum starting from where you released it and ending up at that same point.)
3. Place five washers on the paper clip, and time five complete swings. Now that you are in the swing of things, try 10 washers. Try to pull the washers back the same distance each time.
4. Next, measure the length of the pendulum from the middle of the washers to where the string is attached to the table. Again with 10 washers on the string, time five complete swings.
5. Shorten the pendulum to about half its original length. Time five complete swings. Now shorten the pendulum to about one-fourth its original length. Time five complete swings.

What's Going On Here

When you initially release the washers, gravity pulls them downward, increasing their speed. The washers have energy due to this speed and are able to use the energy to rise to almost the same height on the other side of the swing. The swinging back and forth is an excellent example of how energy is changed from one kind (kinetic) to another kind (potential). The time it takes to complete a full swing back and forth does not depend on the mass of the pendulum. The time for a swing does depend on the length of the pendulum, however. In fact, the time will be halved when the pendulum is one-fourth its initial length.

Try a solid pendulum made from a ruler with small holes. Tape a large paper clip to the table, and bend the paper clip so that the ruler can swing freely on it. No strings attached here!

Try to make a pendulum that takes 1 second to make one complete swing.

Change how far back you pull the washers, and time five complete swings.

SWIFT SWINGER

What You Want to Know

Does the time it takes for a pendulum to make a complete swing back and forth depend on the weight of the swinging object? Does it depend on the length of the string?

What You Think Will Happen

With more weight on a pendulum, the time for a complete swing

        a. will be longer.

        b. will be shorter.

        c. will stay about the same.

With the pendulum's string shorter, the time for a complete swing

        a. will be longer.

        b. will be shorter.

        c. will stay about the same.

What Happened

Record your observations in the tables. (To calculate the time for one complete swing, divide by 5 the time it took for five swings.)

Number of washers	Time for 5 swings	Time for 1 swing
1		
5		
10		

Length of pendulum	Time for 5 swings	Time for 1 swing

What It Means

What can you now say about the effect of a pendulum's weight on the time it takes for the pendulum to make one complete swing? What about the effect of the pendulum's length?

PENCIL PENDULUM

SCIENCE: Energy can be transferred from one swinging pendulum to another.

STUFF: String (40 inches long), ruler, scissors, two washers of the same size, pencil, tape

What to Do

1. Cut two 12-inch strings and two 8-inch strings.
2. Tie each 12-inch string to a washer.
3. Tie the other end of each 12-inch string to a pencil, about 2 inches in from each end of the pencil.
4. Adjust the strings so that when the washers hang freely from the pencil, they hang at the same length. (You can do this by wrapping the longer string around the pencil and holding it in place with a small piece of tape.)
5. Tie the 8-inch strings about $\frac{1}{2}$ inch from each end of the pencil.

6. Tape the 8-inch strings to a tabletop near the edge so that the pencil and washers hang freely and will not hit anything when you start them swinging. Make sure that the pencil is level by adjusting one of the 8-inch strings.
7. Be sure that the pencil and washers are still. Hold one of the washers steady with one hand, and pull the other washer back toward you with the other hand. Let go of both washers, and observe what happens.
8. Remove the 8-inch strings, and tape the pencil to the edge of the tabletop. Repeat Step 7.

What's Going On Here

When the first washer starts swinging, it has energy. This energy is transferred to the pencil, which then transfers it to the other washer. The first washer stops swinging, and the second washer starts swinging. Then the second washer transfers energy back to the first washer. This process continues until friction and air resistance cause both washers to stop swinging. The washers are able to transfer energy back and forth so long as the pencil is free to swing. When the pencil is attached directly to the desk, the washers cannot transfer energy back and forth; the first washer swings by itself until friction and air resistance cause it to stop swinging.

Try It!

Try using a longer string on one of the washers so that the washers hang at noticeably different lengths.

Try hanging three washers, each on the same length string.

Try hanging three washers, two on the same length string and one on a noticeably different length string.

PENCIL PENDULUM

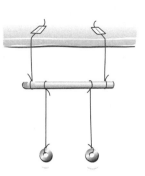

What You Want to Know

Two strings with washers on them are hanging from
a horizontal pencil hanging by two other strings
from a table edge. What happens when you make
just one of the washers swing back and forth?

What You Think Will Happen

When you swing just one of the washers on a hanging pencil,
    a. only that washer will swing, and it will eventually stop.
    b. that washer will swing for a while and then stop, and then
       the other washer will start swinging.
    c. that washer will start the other washer swinging
       immediately, and they will swing together.

When you swing just one of the washers on a pencil attached directly to
a table,
    a. only that washer will swing, and it will eventually stop.
    b. that washer will swing for a while and then stop, and
       then the other washer will start swinging.
    c. that washer will start the other washer swinging
       immediately, and they will swing together.

What Happened

What happened when you swung just one of the washers attached to
the hanging pencil?

_____

What happened when you swung just one of the washers attached to
the pencil taped to the table?

_____

What It Means

What can you now say about how swinging one washer attached to a
hanging pencil affects the other washer's motion?

_____

What can you now say about how swinging one washer attached to a
pencil taped to a table affects the other washer's motion?

_____

HOMEMADE HOVERCRAFT

SCIENCE: Friction can be reduced by moving an object on a cushion of air.

STUFF: Glue gun and glue, thread spool, small plastic plate with smooth bottom, scissors with pointed tip, balloon

What to Do

1. Glue a spool to the middle of the top side of a plastic plate.
2. Punch a hole in the bottom of the plate, using the pointed end of a pair of scissors. Punch the hole from the bottom of the plate into the hole in the spool, and make sure you don't have any rough edges around the hole that would keep the plate from lying flat on the table.
3. Place the plate flat on the table, and give it a little push. Notice how fast it moves and how far it goes.
4. Blow up a balloon and pinch it closed so that no air escapes while you attach the balloon to the top of the spool.

5. Place the plate flat on the table, and unpinch the balloon. Immediately give the plate a little push. Notice how fast it moves and how far it goes.
6. You may have to try a few times before you get the hovercraft to work properly. If it doesn't work at first, try another balloon, make sure the spool doesn't have any holes in the top besides the hole in the center where the air goes through, and make sure the bottom of the plate is smooth and clean.

What's Going On Here

When you push the plate without the balloon attached, it moves a little distance on the table and then stops. Friction between the table and the plate opposes the motion of the plate on the table and causes it to slow down and stop. When the balloon is used, the air that is leaving the balloon forms a thin layer between the table and the plate, reducing friction between the two. The hovercraft rides smoothly along on its cushion of air.

Try It!

Try using a discarded phonograph record or compact disc instead of a plate.

Try different sizes of balloons.

Try different sizes of plates.

Try placing wax paper on the bottom of the hovercraft or on the table.

HOMEMADE HOVERCRAFT

What You Want to Know
For a plastic plate moving across a table, what is the effect of adding a cushion of air between the plate and the table?

What You Think Will Happen
A plastic plate is pushed across a table with and without a cushion of air. With the cushion of air, the plate will
        a. move faster.
        b. move farther.
        c. move faster and farther.
        d. move neither faster nor farther.

What Happened
What happened when you pushed the plastic plate without the cushion of air?

_____

What happened when you pushed the plastic plate with the cushion of air?

_____

What It Means
What do your observations tell you about how a cushion of air affects the movement of a plastic plate on a flat surface?

_____

_____

What changes could you make in your hovercraft design to make it go even faster and farther?

_____

_____

FINICKY FRICTION

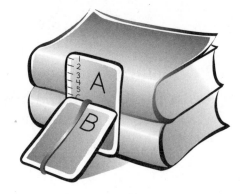

SCIENCE: The force of friction can compete with the force of gravity.

STUFF: Two playing cards, marker, ruler, masking tape, several books, two rubber bands, two pipe cleaners

What to Do

1. Label one playing card "A" and another card "B." Make a mark on card A every quarter inch. Label the marks "1," "2," and so on.
2. Tape card A to a stack of books so that it is upright and flat against the books and so that mark 1 is at the top of the card.
3. Lean the top edge of card B against mark 1 on card A. Move card B to the next lower mark on card A. Continue doing this until card B slips and falls. Record the mark at which card B starts to fall.

4. Wrap a rubber band around card B the long way. Repeat Step 3.
5. Keep the rubber band around card B. Wrap a rubber band around card A the long way, reattach the card to the books, and repeat Step 3.
6. Remove the rubber bands. Wrap a pipe cleaner around card B the long way, and repeat Step 3.
7. Keep the pipe cleaner around card B. Wrap a pipe cleaner around card A, reattach the card to the books, and repeat Step 3.

What's Going On Here

There are two forces acting on card B—the force of gravity trying to pull the card down and the force of friction between the two cards and between card B and the table. The two forces compete so that if the force due to friction is greater than the force of gravity, card B sticks to card A. If the force of friction is smaller than the force of gravity, card B will fall. Initially the force of friction is able to overcome the force of gravity, and card B sticks to card A. But as card B is moved down card A, eventually gravity wins as the force of friction becomes smaller and smaller. Adding the rubber bands or pipe cleaners changes the force of friction so that card B tumbles at a different mark.

Try It!

Try using large cards.

Try using different materials on the cards to increase or decrease friction.

Try placing wax paper or sandpaper on the table where the bottom of card B rests.

See how many cards you can stack against mark 1 on card A without the cards falling. Stack as many cards as you can against the other marks on card A.

FINICKY FRICTION

What You Want to Know

How can you keep a playing card from falling down when it is propped up against another playing card?

What You Think Will Happen

In this activity when playing card B is propped up against card A, card B will start to fall

  a. at the uppermost mark on card A.
  b. at the lowest mark on card A.
  c. somewhere in between.

Card B will rest lowest on card A without falling when

  a. nothing is on card A or card B.
  b. a rubber band is on card B.
  c. a rubber band is on card A and card B.
  d. a pipe cleaner is on card B.
  e. a pipe cleaner is on card A and card B.

What Happened

Record your answers in the table.

Card A	Card B	Mark at which card B fell
Nothing on it	Nothing on it	
Nothing on it	Rubber band on it	
Rubber band on it	Rubber band on it	
Nothing on it	Pipe cleaner on it	
Pipe cleaner on it	Pipe cleaner on it	

In which combination did card B rest lowest on card A without falling?

_____

What It Means

What can you now say about what keeps a playing card from falling when it is propped up against another playing card?

_____

SOLAR SYSTEM SIMULATOR

SCIENCE: An object moving in a circle increases its speed as it gets closer to the center of the circle.

STUFF: Stiff construction paper (12 inches x 9 inches), ruler, scissors, tape, medium-sized plastic cup, marble

What to Do

1. On construction paper, draw a circle with a *diameter* (distance across the circle through the center) of 9 inches. Cut the circle out.
2. Cut a line from the edge of the circle to the center. Form the circle into a cone that will fit into a plastic cup, and tape the cone together.
3. Place the cone with its pointed end down in the plastic cup. Use tape to hold the cone in place in the cup.

4. Place the cup and cone on a table, and hold it with one hand. With the other hand, give a marble a push so that it travels in a circle near the top of the cone.
5. Observe the speed of the marble as it gets closer to the bottom of the cone.

What's Going On Here

When you roll the marble into the cone near the top, you give it a certain starting speed and it moves in a circle at that speed. As the marble moves downward in the cone, the diameter of the circle the marble moves in decreases. As the diameter of the circle decreases, the speed of the marble increases. The same thing happens with the planets going around the sun. A planet close to the sun has a greater speed than a planet that is farther from the sun. In our solar system, Mercury is closest to the sun and moves fastest around it; most of the time, Pluto is farthest away and moves the most slowly.

Try making a larger cone using a larger piece of construction paper.

Try making a cone with more slope or less slope, and notice the difference in the motion of the marble.

Try rolling other objects in the cone. End with a penny on its edge.

SOLAR SYSTEM SIMULATOR

What You Want to Know

What happens to the speed of a marble as it rolls inside a cone in a circular path from the top (wide end) to the bottom (pointed end)?

What You Think Will Happen

As a marble moves in a circular path from the top of a cone to the bottom,

> a. its speed gets greater.
> b. its speed gets smaller.
> c. its speed does not change.
> d. _____.

What Happened

What happened to the speed of the marble as it rolled in a circular path from the top of the cone to the bottom of the cone?

_____

What It Means

What do your observations tell you about how the speed of a marble changes as it moves in a circular path from the top of a cone to the bottom of the cone?

_____

When the marble is at the top of the cone, it is moving in a circle with a large diameter (distance across the circle through the center). When the marble is near the bottom of the cone, it is moving in a circle with a small diameter. What do your observations tell you about how the speed of the marble changes as the diameter of its motion gets smaller?

_____

If you liken the motion of the marble to the motion of planets in our solar system, which planet moves fastest around the sun? Which planet moves the most slowly? Explain your answer.

_____

_____

REVVED-UP REVOLVING RING

SCIENCE: The centripetal force of an object can lift another object vertically.

STUFF: Paper punch, paper cup (3 ounces), pipe cleaner, one Lifesavers candy, kite string (36 inches long), empty pen cartridge, 30 pennies, stopwatch

What to Do

1. Punch two holes across from each other near the top edge of a paper cup. Put a pipe cleaner through the holes, and twist it to form a handle for the cup.
2. Tie a Lifesavers candy to one end of a string.
3. Thread the other end of the string through an empty pen cartridge so that the end of the string with the candy is close to the tapered end of the cartridge.
4. Tie the loose end of the string to the cup handle. Place 10 pennies in the cup.
5. Make sure you have plenty of room around you. Holding the pen cartridge with one hand vertically above your head, twirl the candy in a horizontal circle. Increase the speed of the candy by twirling it faster until the cup begins to rise. Make sure you are holding only the pen cartridge and *not* the string. It will help to support the cup with your other hand until it starts to rise.
6. Time how long it takes for the candy to make five *revolutions* (complete trips in a circle) when the cup's handle is 1 inch below the bottom of the pen cartridge.
7. Place 10 more pennies in the cup, and repeat Step 5.
8. Place 10 more pennies in the cup, and repeat Step 5.

What's Going On Here

An object moving in a circle always has a force acting on it called centripetal, or center seeking, force. The faster the object moves, the greater is the centripetal force. The centripetal force on an object moving in a circle keeps it from going off in a straight line (which it would like to do in accordance with Newton's first law of motion). The candy is able to move in a circle because a force is constantly pulling it inward. That force is due to the weight of the pennies. But as the pennies pull on the candy, the candy pulls back on the pennies (according to Newton's third law of motion). The faster the candy moves, the more it pulls on the pennies, and eventually it is able to lift them.

Try It!

Try different lengths of string.

Try tying more Lifesavers candies to the string.

Name _____ Date _____

REVVED-UP REVOLVING RING

What You Want to Know

When an object is twirled around in a circle, does the speed of the object affect how much force (push or pull) it has?

What You Think Will Happen

As a Lifesavers candy attached to a string is twirled around in a circle, it is able to lift a cup of pennies on the other end of the string. To lift the cup if more pennies are added, the candy will have to

   a. move faster.
   b. move slower.
   c. move at the same speed as before.

What Happened

Record your observations in the table. For each trial, also enter the time it took for one revolution. To calculate that time, divide by 5 the time it took for five revolutions.

Number of pennies in cup	Time for 5 revolutions	Time for 1 revolution
10		
20		
30		

What did you notice about how the *time it took for a revolution* of the candy changed when you added more pennies to the cup?

_____

What did you notice about how the *speed* of the candy changed when you added more pennies to the cup?

_____

What It Means

What do your observations tell you about how the speed of an object moving in a circle affects how much force it has (how much weight it can pull up)?

_____

SPINNING SPEED

SCIENCE: Momentum is conserved as an object moves in a circle.

STUFF: Kite string (36 inches long), one Lifesavers candy, empty pen cartridge, button

What to Do

1. Tie a Lifesavers candy to one end of a string.
2. Thread the other end of the string through an empty pen cartridge so that the end of the string with the candy is close to the tapered end of the cartridge.
3. Tie a button on the loose end of the string.

4. Holding the pen cartridge vertically with one hand above your head, twirl the candy in a horizontal circle until the button moves up against the bottom of the pen cartridge.
5. Stop moving your hand that is holding the pen cartridge, and immediately pull the button downward with your other hand.

What's Going On Here

Physics is full of conservation laws, and this is an example of one of them—the law of conservation of angular momentum. *Angular momentum* is the momentum of an object moving in a circle. Conserving it means that it can't change. The angular momentum of the candy must be the same when it is spinning far from the top of the cartridge and when it is spinning right next to the cartridge. Angular momentum is equal to the product of the mass of the object, its speed, and the radius of the motion. Multiplying those three quantities together must give the same number

regardless of where the object is spinning. Since the mass of the candy doesn't change (provided no one has gotten hungry since the activity started), it is only the product of the speed and the radius that needs to be considered here. The radius times the speed must be constant. So if the radius decreases, the speed of the candy must increase; if the radius increases, the speed of the candy must decrease. In our solar system, Mercury's path around the sun has the smallest radius, and so it moves fastest around the sun.

Try It!

Try spinning the candy in a vertical circle by holding the pen horizontal, and repeat the activity.

Try pulling the button down and then letting it go back up.

Go outside, and try the activity without the button on the end of the string. Get the candy spinning while holding the loose end of the string with one hand. Then let the string go, and notice the path of the candy.

SPINNING SPEED

What You Want to Know

When an object is moving in a circle, what happens to its speed if it moves in a smaller circle?

What You Think Will Happen

In this activity when you pull the button downward, the candy will move in a smaller circle and the speed of the candy will

    a. get larger.
    b. get smaller.
    c. not change at all.

When you pull downward on the button, what changes is

    a. the size of the circle in which the candy moves.
    b. the speed of the candy.
    c. both **a** and **b**.

What Happened

What happened to the speed of the candy when you pulled downward on the button?

_____

When you pulled downward on the button, what changed?

_____

What It Means

What can you now say about how the speed of an object changes as the circle it moves in becomes smaller?

_____

If you liken the motion of the candy to the motion of planets in our solar system, which planet moves fastest around the sun? What observations that you made with the candy lead you to that conclusion?

_____

_____

CIRCULAR COURSE

SCIENCE: A force is necessary for an object to move in a circle. Without a force, the object moves in a straight line.

STUFF: Strip of paper (1 inch x 8 inches), pencil, tape, paper plate (with a smooth surface and a raised lip around the edge), scissors, marble

What to Do

1. Wrap an 8-inch-long strip of paper around a pencil. Use a piece of tape to hold the shape of the resulting 1-inch-tall cylinder. This is the target.
2. Cut a "piece of pie" out of a paper plate equal to about one-fourth of the plate. Throw the small piece away.
3. Secure the plate to a table using a rolled-up piece of tape.
4. Place the target on the table at least 6 inches away from the open edge of the plate at the spot where you predict a marble will hit it after rolling around the plate and then exiting the plate at the opening.
5. Roll the marble along the inside edge of the plate. See what happens to the target. Roll it smoothly, without giving it any bounces or spins.

What's Going On Here

Newton's first law of motion states that an object at rest or in motion in a straight line at a constant speed will remain so unless acted on by a force. The marble rolls in a circle around the inside edge of the plate because the lip of the plate is pushing the marble toward the center of the plate. When it leaves the plate, that force is no longer acting, so the marble moves in a straight line.

Try It!

Make nine targets, and label them "1," "2," "3," and so on. Place them at least 6 inches from the plate in any pattern. Roll the marble, and note the first two targets that are knocked down. Multiply those two numbers together. Set the targets back up, and roll again. Add the products from five rolls of the marble. Try to get the highest score.

There are endless variations on the math game described above. Make up your own!

CIRCULAR COURSE

What You Want to Know

What path does a marble that is moving around a circular plate take when it leaves the plate?

What You Think Will Happen

A marble is moving around the inside edge of a paper plate. When the marble leaves the plate, it will travel along

     a. path 1.
     b. path 2.
     c. path 3.

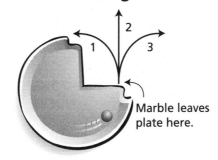

Marble leaves plate here.

What Happened

Draw the path that the marble took when it left the plate.

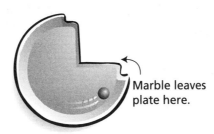

Marble leaves plate here.

What It Means

What do your observations tell you about the path an object takes after it leaves its circular path inside the plate?

_____

_____

MOMENTUM MARBLES

SCIENCE: When one object collides with another object and stops, the object it hits moves.

STUFF: Plastic 12-inch ruler with groove down middle, six small marbles, one large marble

What to Do

1. Place a ruler on a flat surface such as a table.
2. Put five small marbles in the groove of the ruler so that they are touching each other and are located at about the 6-inch mark on the ruler.
3. Place a sixth small marble in the groove of the ruler about an inch away from the other marbles.
4. Hit the sixth marble with your finger so that it rolls toward the other five marbles.
5. Using four small marbles at about the halfway point on the ruler and two small marbles about an inch away, repeat Steps 2–4.
6. Using five small marbles at about the halfway point on the ruler and one large marble about an inch away, repeat Steps 2–4.

What's Going On Here

When the sixth marble is rolled toward the other marbles, it has momentum due to its mass and speed. A very important law in physics (the law of conservation of momentum) states that the momentum before the collision has to be the same as the momentum after the collision. Before the collision, only the sixth marble is moving. After the collision, only one of the marbles from the group that was hit is moving, and it is moving at almost the same speed as the sixth marble originally was. So the momentum after the collision is almost the same as the momentum before the collision. Actually the speed of the marble moving after the collision is slightly less than the original speed of the sixth marble. This is because some energy has been converted to sound and heat energy during the collision itself. You can hear the sound energy as the marbles hit each other. They also heat up slightly as they rub against each other during the collision.

Try It!

Try using four still marbles, three still marbles, two still marbles, and then one still marble on the ruler.

Try using large marbles for the group of five and a smaller marble as the sixth one.

Trying propping one grooved ruler up at a slope so that it feeds into a second ruler that lies flat. Place five marbles on the flat ruler at the end where the sloped ruler feeds in. Roll one marble down the slope. Then roll two marbles down the slope together.

MOMENTUM MARBLES

What You Want to Know

What happens when you roll a marble so that it hits the end of a line of five other marbles that are all touching each other?

What You Think Will Happen

You roll a sixth marble along in a groove containing a line of five marbles all touching each other. When the sixth marble hits the end of the line of marbles,

      a. the sixth one stops, and the one farthest from it moves.
      b. the sixth one keeps moving, and so do all the others.
      c. the sixth one stops, and the other five move.

When two marbles hit the end of a line of four other marbles,

      a. the first two stop, and the other four move.
      b. the first two keep moving, along with the other four.
      c. the first two stop, and the two farthest from them move.

When one large marble hits the end of a line of five small marbles,

      a. the large marble stops, and two small ones move.
      b. the large marble keeps moving, along with the other five.
      c. the large marble stops, and one small one moves even faster than the large one.

What Happened

What happened when you rolled one marble toward five marbles in the groove?

_____

What happened when you rolled two marbles toward four marbles?

_____

What happened when you rolled one large marble toward five small marbles?

_____

What It Means

What can you now say about what happens when you roll a marble at a line of marbles that are touching each other?

_____

PEOPLE-POWER PROJECTILES

SCIENCE: A paper rocket can be launched using the force of moving air.

STUFF: Strip of paper (6 inches x 3 inches), pencil, tape, straw, scissors, tape measure

What to Do

1. Roll a paper strip around a pencil, and tape it near the middle to hold it in place. Slide the resulting 6-inch-tall cylinder down so that about an inch is off the pencil.
2. Crimp the free end of the cylinder to make the pointed nose of a rocket, and tape it in place.
3. Take the rocket off the pencil, and put it on a straw. Blow through the straw to launch the rocket. Be sure to aim the rocket away from other people.
4. Attach fins to the rocket by cutting triangles of paper and taping them near the open end of the rocket.
5. Launch the rocket again by blowing through the tube.
6. Launch the rocket at different angles (vertical, horizontal, and halfway between) three times each, and measure how far it travels each time. Try to blow through the straw with the same strength each time.

What's Going On Here

When you blow air through the straw, the air pushes against the tip of the rocket, causing it to move. Newton's third law of motion states that for every action there is an equal (in size) and opposite (in direction) reaction. Forces always occur in pairs. In this case, moving air pushes against the tip of the rocket; the rocket in turn pushes back against the moving air. This is what causes the rocket to move forward.

Try It!

Try different fin designs.

Try attaching a different number of fins.

Try making a smaller or larger rocket.

Try putting the straw into the tube different distances before launching it.

Play a game with your paper rocket by making a large circular target on the floor and trying to hit it with the rocket.

Try launching the rocket at many different angles, and measuring how far it travels each time.

PEOPLE-POWER PROJECTILES

What You Want to Know

What happens when you blow through a straw into a tube of paper shaped like a rocket? At what angle should you launch the rocket to make it go farthest?

What You Think Will Happen

When you blow through a straw to launch a paper rocket, it will travel

    a. farthest when you hold it vertically.
    b. farthest when you hold it horizontally.
    c. farthest when you hold it halfway between vertical and horizontal.
    d. about the same distance at all angles.

What Happened

Record your observations in the table. (To calculate the average distance, add the three distances for the angle and divide that number by 3.)

Angle of launch	Distance traveled (1st time)	Distance traveled (2nd time)	Distance traveled (3rd time)	Average distance traveled
Vertical				
Horizontal				
Halfway between				

At which angle did the rocket travel the farthest average distance?

_____

How did the fins affect the motion of the rocket?

_____

What It Means

What can you now say about how the angle of launch for a paper rocket affects the distance it travels?

_____

How could you redesign the paper rocket to travel even farther?

_____

© Dale Seymour Publications®

LOOP LAUNCHER

SCIENCE: A rocket moves in accordance with Newton's third law of motion.

STUFF: Two straws (each with a different size of opening), piece of paper ($8\frac{1}{2}$ inches x 11 inches), ruler, scissors, tape, clay, small plastic soft drink bottle

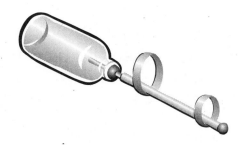

What to Do

1. Try to throw the larger-opening straw so that it glides smoothly through the air.
2. Cut one strip of paper $\frac{1}{2}$ inch by 5 inches and another strip $\frac{1}{2}$ inch by 8 inches. Tape each strip into a loop.
3. Tape one loop on the larger-opening straw about an inch from one end. Tape the other loop on the same straw about an inch from the other end. Both pieces of tape should be on the same side of the straw so that the loops line up when you look through them.
4. Launch the straw rocket into the air by gently throwing it with the taped edge downward and the smaller loop facing away from you. Try to launch the rocket so that it glides smoothly through the air.

5. Use a little clay to plug the straw rocket on the end with the small loop. Try throwing the straw rocket again.
6. Put the smaller-opening straw in a plastic bottle with about 4 inches of straw sticking out of the bottle. Put the clay around the bottle opening to hold the straw in place.
7. Push the small straw in the plastic bottle into the straw rocket. Be sure the large straw of the rocket does not press into the clay in the bottle opening.
8. Aim the rocket up and away from you and other people. Quickly and firmly squeeze the bottle. It will take a little practice and patience, but soon you will be an expert straw rocket launcher.

What's Going On Here

When you squeeze the bottle, you are increasing the air pressure inside the bottle by making the air particles get closer together. The air pressure builds up inside the large straw until it pushes out the back opening, propelling the large straw through the air. This is a good example of Newton's third law of motion (for every action, there is an equal and opposite reaction). The air from the straw rocket pushes out in one direction, and the rocket moves in the other direction.

Try It!

Try using different sizes of loops on the straw rocket.

Try making a bigger rocket and using a bigger soft drink bottle.

Make a target, place it on the floor, and try to hit it by launching the straw rocket.

Try launching the rocket without putting clay in the end of the large straw.

LOOP LAUNCHER

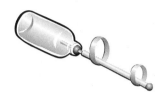

What You Want to Know

How does a rocket propel itself through the air?
How does it need to be designed so that it can move
smoothly through the air at the greatest speed?

What You Think Will Happen

Design features that will cause a straw rocket to move smoothly
through the air include

        a. attaching loops to the front and back of the rocket.
        b. placing clay in the front opening of the rocket.
        c. both **a** and **b**.
        d. neither **a** nor **b**.

What Happened

Describe the motion of the straw when you launched it by throwing it.

_____

Describe the motion of the straw rocket with paper loops when you
launched it by throwing it.

_____

Describe the motion of the straw rocket with paper loops and clay plug
when you launched it by throwing it.

_____

Describe the motion of the straw rocket with paper loops and clay plug
when you launched it using the plastic bottle.

_____

What It Means

What do your observations tell you about how a rocket propels itself
through the air?

_____

How could you redesign the rocket and launcher to make the rocket go
even farther?

_____

PINE PADDLEBOAT

SCIENCE: A wound-up rubber band has potential energy. When the rubber band unwinds, the potential energy can be changed into kinetic energy.

STUFF: Piece of pine wood (4 inches x $3\frac{3}{4}$ inches x $\frac{3}{4}$ inch), sandpaper, pine molding (7 inches x $\frac{3}{4}$ inch x $\frac{1}{4}$ inch), three rubber bands, pine lath (3 inches x $1\frac{3}{4}$ inches x $\frac{1}{4}$ inch), tub of water

What to Do

1. Use a piece of pine wood as the body for your paddleboat. If the edges of the wood are rough, sand them.
2. Attach side supports (pine molding) to the boat body, using two rubber bands. The door molding should be flat against the 4-inch sides of the boat and should be flush at the front end of the boat.
3. Stretch a third rubber band around the side supports at the back end of the boat.
4. Put the paddle (pine lattice) through the back rubber band going lengthwise (the rubber band will be crossing the paddle's short side at the middle). Wind the paddle up.
5. Place the boat in a tub of water, and watch it go!
6. Experiment with which way you need to wind the rubber band to make the boat go in the forward direction.
7. Place the paddle through the back rubber band going crosswise (the rubber band will line up with the middle of the paddle's long side). Wind it up, and let the boat go.

What's Going On Here

When you wind the paddle up, you are storing potential energy in the rubber band. The boat moves when you place it in the water because the potential energy is changed to kinetic energy, or energy of motion, as the rubber band unwinds.

Try It!

Try making the front of the boat triangular so that it moves through the water more easily.

Try another shape for the boat.

Try a wider or narrower paddle.

Add lightweight materials to decorate your boat or to make it look more realistic.

PINE PADDLEBOAT

What You Want to Know

How does the size of the paddle on a paddleboat affect how the boat moves in water? Which way do you have to wind the rubber band on the boat to make it move forward ?

What You Think Will Happen

When a paddle is put on a boat so that the narrow side of the paddle pushes the water,

  a. the boat will move slower than when the wide side pushes.
  b. the boat will move faster than when the wide side pushes.
  c. the boat will move at about the same speed as when the wide side pushes the water.

To make the boat move in a forward direction, you should wind the paddle

  a. toward the front of the boat.
  b. toward the back of the boat.
  c. either direction.

What Happened

What happened when the narrow side of the paddle pushed the water?

_____

Which way did you have to wind the paddle for the boat to move forward in the water?

_____

What happened when the wide side of the paddle pushed the water?

_____

What It Means

What can you now say about how the way the paddle is positioned affects the way the boat moves through the water?

_____

What do your observations tell you about how to wind the paddle so that the boat moves in a forward direction?

_____

SIMPLE SCALE

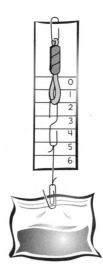

SCIENCE: It takes less force to pull an object up a ramp than it takes to lift the object vertically.

STUFF: Straw, scissors, rubber band (about 2 inches long), three large paper clips, tape, stiff cardboard (3 inches x 11 inches), string (4 inches long), ruler, marker, sealable plastic sandwich bag, sand, wooden board (about 2 feet long), several books

What to Do

1. Cut a straw to a length slightly shorter than the length of a rubber band.
2. Put a paper clip on the cardboard in the middle of one of the 3-inch sides. Loop one end of the rubber band on the paper clip.
3. Pull the rubber band through the piece of straw, and tape the straw to the cardboard so that it is just touching the paper clip.
4. Unbend another paper clip, and attach it to the free end of the rubber band.
5. Tie one end of a string to the second paper clip and the other end to a third paper clip.
6. Make horizontal marks on the cardboard every inch, starting at the bottom of the straw and going down. From the straw, label the marks "0," "1," "2," and so on.
7. Fill a sandwich bag halfway up with sand, and close the bag. Open the bot-

tom paper clip on the scale a little, and stick it through the top edge of the sand bag. (Be careful not to make a hole big enough to spill the sand.) Holding the scale at the top, lift the sand bag 6 inches into the air. Be sure to keep the scale and sand bag vertical. Record the position of the top part of the second paper clip. This number corresponds to the force that is needed to lift the sand bag.

8. Prop a board up on several books so that the top end of the board is 6 inches above a table surface. Place the sand bag at the bottom of the inclined board, and pull it up the board by holding the top of the scale. Record the position of the top part of the second paper clip as you pull the sand bag up the board.
9. Repeat Step 8 with the board flat on the horizontal table surface.

What's Going On Here

It takes more force to lift an object vertically than to pull the same object up a ramp. Even so, when the object is pulled up the incline, the force of friction makes you have to pull a little harder than you would otherwise have to. If you place some slip-

pery material on the board, it will take even less force to pull the object up the ramp. Pulling the object along a horizontal surface takes even less force than pulling it up a ramp.

SIMPLE SCALE

What You Want to Know

Is it easier to lift an object a certain height straight up or to pull it up a ramp to that height? Is it easier to lift an object vertically or to move it the same distance on a horizontal surface?

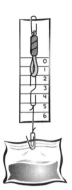

What You Think Will Happen

Moving an object to a certain height will be
- a. easier if you lift it vertically.
- b. easier if you pull it up a ramp.
- c. equally easy either way.

If a sand bag is to be moved on a horizontal surface the same distance that it was lifted vertically,
- a. more effort will be needed.
- b. the same effort will be needed.
- c. less effort will be needed.

What Happened

Record your observations in the table.

Sand bag moved 6 inches	Force needed (number from scale)
Vertically	
Up a ramp	
Horizontally	

What It Means

What can you now say about which takes more effort, lifting an object vertically or moving it to the same height using a ramp?

_____

What can you now say about which takes more effort, lifting an object vertically or moving it the same distance on a horizontal surface?

_____

What could you do to make it easier to move an object on a horizontal surface?

_____

CHAPTER 2 ❚ HEAT

Highlighted Heat Happenings

- Heat is energy due to motion. When an object is heated, molecules in the object move faster. When an object is cooled, the molecules move slower.
- As the temperature of a gas (such as air) increases, the molecules move faster and the gas takes up more space—that is, a greater volume.
- As the temperature of a gas decreases, the molecules slow down and the gas takes up less space—in other words, a smaller volume.
- Because hot air is less dense than cold air, hot air will rise.
- Hot water is less dense than cold water and will rise above cold water. But something unusual happens in water close to its freezing point that causes the colder water to rise.
- Solids, liquids, and gases expand when they are heated because their molecules are moving faster and take up more space. There are some exceptions; rubber, for example, contracts when it is heated.
- Solids, liquids, and gases contract when they are cooled because their molecules are moving slower and take up less space. There are some exceptions; water, for example, expands when it is about at its freezing temperature.
- Saltwater has a lower freezing temperature than plain water. That is why salt is sometimes used to melt ice on roads and sidewalks. It is also why salt is used to make ice cream in one of the activities in this chapter.

CHILLY CHALLENGE

SCIENCE: Hot air takes up more space than cold air.

STUFF: Two large bowls; ice; hot and cold tap water; large, round balloon; plastic soft drink bottle

What to Do

1. Inflate a balloon and let the air out a few times so that it's easy for the balloon to inflate later when it's put on a bottle.
2. Fill one bowl about one-fourth full of ice. Add cold water to make it half full.
3. Fill another bowl one-half full of hot water.
4. Put the deflated balloon over the mouth of a soft drink bottle.
5. Place the soft drink bottle in the bowl of hot water. Observe for several minutes.
6. Take the bottle out of the hot water, and immediately place it in the bowl of ice water. Observe for several minutes.

What's Going On Here

When air is heated, the air particles, or molecules, move faster and take up more space. Placing the soft drink bottle in the hot water causes the air inside it to expand. There are no more molecules of air in the bottle than before it was put in the hot water, but they are moving faster now, are farther apart, and take up more space, causing the balloon to inflate. When the soft drink bottle is placed in the ice water, the opposite happens. The same number of molecules as before get closer together, move more slowly, and take up less space, so the balloon deflates.

Try It!

Try different sizes of balloons.

Try different sizes of soft drink bottles.

Place the soft drink bottle in the ice water first and then in the hot water.

Use a piece of string to measure how big the balloon becomes in the hot water. Change the temperature of the water, and measure the size of the balloon again.

Instead of putting the bottle in hot water, place it in the window on a sunny day with a piece of black construction paper wrapped around the bottle.

CHILLY CHALLENGE

What You Want to Know

A balloon is put over the mouth of a bottle. What happens to the balloon when the bottom part of the bottle is placed in hot water? What happens when it is placed in ice water?

What You Think Will Happen

When the bottom of a bottle that has a balloon over its mouth is placed in hot water,

       a. the balloon will become larger.
       b. the balloon will become smaller.

When the bottom of a bottle with a balloon over its mouth is placed in ice water immediately after it was in hot water,

       a. the balloon will become larger.
       b. the balloon will become smaller.

What Happened

What happened to the balloon when you placed the bottle in hot water?

_____

What happened to the balloon when you placed the bottle in ice water immediately after it was in hot water?

_____

What It Means

What do your observations tell you about what happens to air when it is heated?

_____

What do your observations tell you about what happens to air when it is cooled?

_____

What do your observations tell you about how much space cold air occupies compared with how much space the same amount of hot air occupies?

_____

BALLOON BATH

SCIENCE: Hot air takes up more space than cold air.

STUFF: Large, round balloon; three strings (each 40 inches long); waterproof marker; ruler; two large bowls; ice; hot and cold tap water

What to Do

1. Blow up a balloon and let the air out a few times so that it will inflate more easily later in the activity. Then blow up the balloon until it is half filled with air. Tie the neck of the balloon so that the air stays inside.
2. Wrap a string around the fattest part of the balloon, and tie the string in a bow knot.
3. Use a marker to draw a line around the balloon where the string is tied.
4. Mark the string on each side of the knot so that when you take the string off the balloon, you can measure how much string went around the balloon. Take the string off the balloon, and use a ruler to measure the length of the string between the marks.
5. Fill one bowl about one-fourth full of ice. Add cold water to make it half full.
6. Fill another bowl one-half full of hot water.
7. Place the balloon in the bowl of hot tap water for 1 minute. Roll the balloon around so that it gets a good bath.
8. Quickly take another piece of string, and tie it around the balloon along the line you marked in Step 3. Measure the length of the string as you did in Step 4.
9. Repeat Steps 7 and 8, using the ice water.

What's Going On Here

When the air inside the balloon is heated, there are the same number of air molecules, and they are the same size as before. But in the heated balloon they move faster and take up more space than in the cooler balloon; therefore the distance around (the *circumference* of) the heated balloon is larger. When you place the balloon in ice water, the air molecules slow down and move closer together; the circumference of the balloon decreases.

Try It!

Try taking the balloon outside on a very cold day or putting it in the freezer.

Try placing the balloon in a window on a sunny day or placing it in the window with a dark cloth wrapped around it.

Try using a balloon that is half filled with helium.

Try using a balloon that is half filled with carbon dioxide from the Balloon-Blowing Bottle activity on page 234.

BALLOON BATH

What You Want to Know
What happens to a partly inflated balloon when it gets hot and when it gets cold?

What You Think Will Happen
When a partly inflated balloon gets hot, the balloon
> a. will become larger.
> b. will become smaller.
> c. will stay the same size.

When a partly inflated balloon gets cold, the balloon
> a. will become larger.
> b. will become smaller.
> c. will stay the same size.

What Happened
Record your observations in the table. (The *circumference* of the balloon is the distance around its fattest part.)

Balloon temperature	Balloon circumference
Room temperature	
Hot	
Cold	

What It Means
What can you now say about what happens to the size of a balloon when it is placed in hot water?

_____

What can you now say about what happens to the size of a balloon when it is placed in ice water?

_____

What can you now say about how much space cold air occupies compared with how much space the same amount of hot air occupies?

_____

BROTH BUBBLES

SCIENCE: Hot air takes up more space than cold air.

STUFF: Three medium-sized bowls, bubble solution (purchase, or use recipe from the Bigger, Better Bubbles activity on page 238), hot and cold tap water, ice, empty soup can (10 ounces)

What to Do

1. Carefully clean the inside of an empty soup can.
2. Pour about $\frac{1}{2}$ inch of bubble solution into a bowl that is wide enough so that a soup can is able to sit on end in the solution.
3. Fill the second such bowl about one-fourth full of ice. Add cold water to make it half full.
4. Fill a third such bowl halfway with hot water.
5. Place the soup can in the bowl of ice water for 1 minute. Be sure that its closed bottom half is immersed.

6. Take the can out of the ice water, and immediately put the open end into the bubble solution.
7. Take the can out of the bubble solution, and immediately place the can upright in the hot water so that its bottom half is submerged. Watch the bubble.
8. When the bubble pops, take the can out of the hot water and immediately put the open end back into the bubble solution.
9. Take the can out of the bubble solution, and immediately place the can upright in the ice water so that its bottom half is submerged. Watch the bubble.

What's Going On Here

When you place the can in the ice water initially, the air inside the can cools down and the particles of air move closer together. Moving the can from the ice water to the hot water causes the air inside the can to heat up; the particles of air move faster and take up more room. The bubble on the top of the can expands outward because the air inside the can is heated and needs more room than it had when it was cooler. Moving the can from the hot water to the ice water has the opposite effect. The air particles move closer together, taking up less space, and the bubble slides down the can.

Try It!

Try heating the can up with your hands after it has been in the refrigerator for a few minutes and then placed in the bubble solution.

Try using a can with both ends removed.

Try other can sizes.

BROTH BUBBLES

What You Want to Know

The open end of an empty soup can is placed in a
bubble solution. If the can is cold, what happens to
the bubble when the closed end of the can is placed
in hot water? What happens to the bubble if you use
a hot can and place it in ice water?

What You Think Will Happen

When you place the bottom half of a cold can with a bubble on its top
end into hot water, the bubble

        a. will get a little bigger.

        b. will slide down the inside of the can.

        c. will not change.

When you place the bottom half of a hot can with a bubble on its top
into ice water, the bubble

        a. will get a little bigger.

        b. will slide down the inside of the can.

        c. will not change.

What Happened

What happened to the bubble when you placed the cold can in hot
water?

_____

What happened to the bubble when you placed the can in ice water?

_____

What other observations did you make?

_____

_____

What It Means

What do your observations tell you about the amount of space cold air
takes up compared with the amount of space hot air takes up?

_____

FANTASTIC FIZZING FEAT

SCIENCE: Alka-Seltzer tablets combine with water to produce carbon dioxide. Hot gas takes up more space than cold gas.

STUFF: Clear plastic cup (8 or 10 ounces), hot and cold tap water, aluminum pie pan, waterproof markers (two colors), two Alka-Seltzer tablets, ruler, drying cloth

What to Do

1. Fill a plastic cup with cold water. Invert it carefully on a pie pan. Do this by first inverting the pie pan on top of the cup of water and then flipping the combination as a unit, holding the cup tightly against the pie pan.
2. Place about $\frac{1}{2}$ inch of cold water in the pie pan.
3. Mark on the outside of the cup the level of the water that is *inside the cup.*
4. Predict what will happen when you slide one Alka-Seltzer tablet under the inverted cup. Draw a line on the cup in one color to indicate where you think the final water level inside the cup will be.

5. Carefully slide the tablet under the inverted cup so that no water from the cup escapes. Gently hold the bottom of the cup down so that it doesn't move around in the pie pan as you work.
6. After the tablet dissolves, draw a line on the cup in another color to indicate the actual final water level inside the cup. Measure the distance between the water level line before the tablet was added and the line for the final water level.
7. Dump the water from the cup and pie pan, and thoroughly dry them.
8. Repeat Steps 1–6, using hot water.

What's Going On Here

Alka-Seltzer tablets are a combination of a dry acid and sodium bicarbonate. When the dry acid is mixed with water, it reacts with the sodium bicarbonate to produce carbon dioxide. This is similar to the reaction that takes place when you mix vinegar and baking soda. The carbon dioxide gas takes up space and therefore pushes water out of the cup. A hot gas takes up more space than a cooler one with the same number of molecules. When hot water is used to dissolve the tablet, hot carbon dioxide gas is produced; since it takes up more space than cool carbon dioxide gas, it pushes more water out of the cup.

Try It!

Try to measure the water level in the pan before and after adding the tablet.

Try timing the reaction. How long does it take for the tablet to totally dissolve in cold water? In hot water?

Try other antacid tablets.

FANTASTIC FIZZING FEAT

What You Want to Know

What happens when an Alka-Seltzer tablet is put into an inverted cup of cold water? How are the results different with hot water?

What You Think Will Happen

When an Alka-Seltzer tablet is put into an inverted cup of cold water, it will dissolve and

a. push less than half the water out of the cup.
b. push all the water out of the cup.
c. push more than half the water out of the cup.
d. not push any water out of the cup.

If hot water is used instead of cold water,

a. more water will be pushed out of the cup.
b. less water will be pushed out of the cup.
c. the same amount of water will be pushed out of the cup.

What Happened

What happened when you placed the Alka-Seltzer tablet in the inverted cup of cold water?

_____

What happened when you placed the Alka-Seltzer tablet in the inverted cup of hot water?

_____

What It Means

What can you now say about what happens when an Alka-Seltzer tablet is placed in water?

_____

_____

What can you now say about how the temperature of a gas affects the amount of space the gas takes up?

_____

JOLLY JACK-O'-LANTERN

SCIENCE: Fire needs oxygen to burn. Hot air rises.

STUFF: Knife, pumpkin, large spoon, candle, matches

What to Do

1. Using a knife, cut a large circle around the stem of a pumpkin, and remove the top. Using a spoon, clean out the inside of the pumpkin.
2. Place a candle inside the pumpkin, and light it. Leave the candle burning.
3. Place the top back on the pumpkin, and notice what happens to the candle. After a short time, the flame will go out.
4. Carve the nose of the jack-o'-lantern face, light the candle, and put the top on. Again, after a short time, the flame will go out.
5. Carve the mouth for the jack-o'-lantern, light the candle, and put the top on. This time, the flame will continue to burn.

6. Put the candle flame out. Carve the eyes for the jack-o'-lantern to finish it. Then light the candle again.
7. Light a match, and extinguish it. Quickly place the match near the mouth of the jack-o'-lantern. The smoke from the match will go into the mouth of the jack-o'-lantern.
8. Light a match, and extinguish it. Quickly place the match near the eyes of the jack-o'-lantern. The smoke from the match will go outward away from the eyes.

What's Going On Here

When the pumpkin has no holes in it and the top is on, the candle burns for a short time and then goes out because it uses up all the oxygen inside the pumpkin. When a nose is carved, the candle goes out because the oxygen-depleted air inside the pumpkin has nowhere to escape so that a fresh supply of oxygen from outside the pumpkin can enter. Carving the mouth allows a fresh supply of oxygen in; the oxygen-depleted air can then escape from the nose of the jack-o'-lantern. In the completed jack-o'-lantern, the hot oxygen-depleted gas rises and goes out the eyes. The cooler air from outside the jack-o'-lantern comes in through its mouth. Since air is not visible, you can't actually see where it enters and leaves the jack-o'-lantern. Using the extinguished-match test on the completed jack-o'-lantern shows that air really does enter through the mouth and leave through the eyes.

Try It!

Try using the extinguished-match test near the nose of the jack-o'-lantern.

Try carving one large hole in the pumpkin. (You can use the completed jack-o'-lantern if you plug the eyes, nose, and mouth with the pieces of the pumpkin you cut out.)

JOLLY JACK-O'-LANTERN

What You Want to Know

What happens to the candle flame inside a jack-o'-lantern when you make one part of its face at a time? Where does the heated air from the candle flame go in a completed jack-o'-lantern?

What You Think Will Happen

When you place the top on a hollowed-out pumpkin that has not been carved into a jack-o'-lantern,

a. a lighted candle will go out.
b. a lighted candle will continue to burn.

When you place the top on a jack-o'-lantern with only a nose carved,

a. a lighted candle will go out.
b. a lighted candle will continue to burn.

When you place the top on a jack-o'-lantern with a nose and mouth,

a. a lighted candle will go out.
b. a lighted candle will continue to burn.

What Happened

What happened to the candle flame when you placed the top on a

hollowed-out pumpkin? _____

jack-o'-lantern with only a nose? _____

jack-o'-lantern with a nose and a mouth? _____

Where did the smoke from the extinguished match go when you held it near the

jack-o'-lantern's mouth? _____

jack-o'-lantern's eyes? _____

What It Means

What do your observations tell you about what fire needs to burn?

_____

What do your observations tell you about how hot gases move in a completed jack-o'-lantern?

_____

WHICH WAY, WARM WIND?

SCIENCE: Hot water is less dense than cold water.

STUFF: Five- or 10-gallon fish tank; lukewarm, hot, and cold tap water; two small jars; food coloring (red and blue); colored pencils (red and blue) for student page

What to Do

1. Fill a fish tank almost to the top with lukewarm water. Let it stand for about an hour to come to room temperature.
2. Fill one small jar with hot water, and put in seven drops of red food coloring.
3. Fill another small jar with cold water, and put in seven drops of blue food coloring.

4. Place the hot-water jar on the bottom of the fish tank in a horizontal position, with the opening toward the center of the tank. At the same time, hold the cold-water jar just below the surface of the water in a horizontal position at the other end of the tank, again with the opening toward the center of the tank. Observe what happens.

What's Going On Here

The cold water is denser than the room-temperature water, and it sinks toward the bottom of the tank. The hot water is less dense than the room-temperature water, and it floats toward the surface. The same thing happens with air and causes winds. The earth is heated by the sun, and air close to the earth becomes heated too. This heated air rises, and cooler air rushes in to take its place. The cooler air is then heated, and the cycle repeats itself. The effect is

very noticeable near large bodies of water. By late afternoon, the earth has been heated up quite a bit, and the hot air rises. Cooler air from the large body of water rushes in to take the place of the rising hot air, so we feel the breeze coming from the water. In the morning, the situation is usually reversed. Cooler air from the land rushes in to take the place of the rising warm air from the water surface.

Try It!

Try placing the cold water at the bottom of the tank and holding the hot water just below the surface of the water.

Try using cold water in the fish tank.

Try using hot tap water in the fish tank.

WHICH WAY, WARM WIND?

What You Want to Know

Does cold water sink or rise? Does hot water sink or rise? What does the movement of cold and hot water have to do with wind?

What You Think Will Happen

An open jar of hot water (colored red) is placed horizontally on the bottom of a fish tank filled with room-temperature water. An open jar of cold water (colored blue) is placed horizontally near the top of the tank. As the water from the two jars flows out,

        a. the blue water will rise, and the red water will sink.
        b. the red water will rise, and the blue water will sink.
        c. both the red and the blue water will sink.
        d. both the red and the blue water will rise.

What Happened

Draw lines to show how the water from the two jars flows when the jars were placed in the fish tank. Use a blue pencil for the cold water and a red pencil for the hot water.

Describe what happened when the jars were placed in the water.

_____

What It Means

Density is how much matter there is within a given space. A box filled with feathers is less dense than the same box filled with books. A less-dense liquid will float on top of a denser liquid. What do your observations tell you about which is denser, hot water or cold water?

_____

If you think of the water in this activity as being like air, what can you now say about what causes wind—that is, the movement of air?

_____

© Dale Seymour Publications®

MELTING MANY MASSES

SCIENCE: The density of water affects its ability to melt an ice cube. Denser liquids sink in less-dense liquids.

STUFF: Water, food coloring (blue), ice cube tray, refrigerator (or freezer), three clear plastic cups (8 or 10 ounces), salt, waterproof markers, drying cloth, isopropyl alcohol, colored pencil (blue) for student page

What to Do

1. Several hours before starting this activity, make a dozen ice cubes using water that has been colored with blue food coloring. The water should be deep blue.
2. Fill two plastic cups to the same height with lukewarm water. Dissolve 2 tablespoons of salt in one of the cups. Label the cups "plain water" and "saltwater."
3. Place an ice cube in each cup at the same time. Watch the ice cubes melt, and observe which one completely melts first.
4. Empty, rinse, and dry the cups. Fill two cups to the same height with lukewarm water. Dissolve 2 tablespoons of salt in one of the cups. Fill the third cup to the same height as the other two with isopropyl alcohol. Label each cup as to its contents. Wait for about 20 minutes so that the three cups are all at room temperature.
5. Place an ice cube in each cup. This time, notice where the blue water from the ice cube settles as the ice cube melts in each solution.

What's Going On Here

When you place an ice cube in plain water and in saltwater, the ice cube in the plain water melts first. That's because saltwater is denser than plain water. As water from the ice cube melts in saltwater, it is difficult for it to disperse through the salt solution, so it stays close to the ice cube, insulating it somewhat from further melting. The water melting from the ice cube in plain water, on the other hand, actually sinks away from the ice cube, allowing warmer water to surround the ice cube and speed up the melting process. When you place the three ice cubes in plain water, saltwater, and iso-propyl alcohol, you are able to see how the density of liquids affects how they sink or float. Less-dense liquids will end up on top, and denser ones will end up on the bottom. Saltwater is denser than plain water, so the melted blue water ends up on top in that cup. Isopropyl alcohol is less dense than plain water, so the melted blue water ends up on the bottom of that cup. In the plain-water cup, the melted blue water at first sinks to the bottom because it is colder and denser, but it soon mixes with the rest of the water.

MELTING MANY MASSES

What You Want to Know
Does an ice cube melt faster in plain water or saltwater? If an ice cube melts in a cup of plain water, of saltwater, and of alcohol, where will the melted water end up?

What You Think Will Happen
One ice cube is placed in a cup of lukewarm saltwater, and another one in a cup of lukewarm plain water. The ice cube that will melt first is
        a. the one in plain water.
        b. the one in saltwater.
        c. neither; they will melt at the same rate.

In each drawing below, color in the area where the water from a melted ice cube will end up. Use a blue pencil.

Plain water

Saltwater

Alcohol

What Happened
Did the ice cube melt faster in plain water or saltwater?

_____

In each drawing below, color in the area where the water from the melted ice cube settled. Use a blue pencil.

Plain water

Saltwater

Alcohol

What It Means
Density is how much matter there is within a given space. Density is often referred to as mass per volume. A box filled with feathers is less dense than the same box filled with books. What can you now say about the relative densities of plain water, saltwater, and alcohol?

_____

THRIFTY THERMOMETER

SCIENCE: As air is heated, it expands. The more it expands, the more room it needs.

STUFF: Two medium-sized bowls; lukewarm, hot, and cold tap water; ice; small cup; food coloring; clear straw (available at restaurants); small glass bottle; clay

What to Do

1. Fill one bowl half full with hot tap water.
2. Fill another bowl about one-fourth full of ice; add cold water to make it half full.
3. Fill a small cup with lukewarm water, and add several drops of food coloring to color the water.
4. Place a straw in the cup of water to a depth of about an inch. Put your index finger over the top of the straw, and remove the straw from the water.
5. Keep your index finger over the top of the straw, and place the other end of the straw into a glass bottle. Push the straw about halfway into the bottle, and let a partner mold clay around the top of the bottle, sealing the mouth of the bottle

and holding the straw in place without crushing it.

6. Take your finger off the top of the straw. The water should rise in the straw a little but shouldn't go out the top.
7. Hold your hands around the lower part of the bottle for a few minutes. The water in the straw should rise still farther. Remove your hands before the water flows out the top.
8. Place the bottle in a bowl of hot water. The water will be warmer than your hands, so the water in the straw should rise even more. If the water starts to flow out the top of the straw, quickly move to Step 9.
9. Place the bottle in a bowl of ice water. The water in the straw should go down.

What's Going On Here

As the air inside the bottle is heated by your hands or by the hot water, the air molecules move faster and need to take up more space. There is space available in the straw if the air could just push the water up a bit. It gladly does this! When the bottle is placed in ice water, the opposite thing happens. The air molecules slow down, get closer together, and take up less space. The water in the straw is delighted to take up the space vacated by the air molecules, and it moves down in the straw to do that!

Try It! Try calibrating your straw thermometer by making marks on the straw and comparing the actual temperature of the water in the bowl to the water level in the straw.

Try using a bottle filled to the top with water.

THRIFTY THERMOMETER

What You Want to Know
What happens when air is heated? What happens when air is cooled?

What You Think Will Happen
A straw containing a column of water sticks out the top of a sealed bottle. When the bottle is heated by your hands or by hot water,

      a. the water inside the straw will rise.
      b. the water inside the straw will drop.
      c. the water inside the straw will not move at all.

When the bottle is placed in ice water,

      a. the water inside the straw will rise.
      b. the water inside the straw will drop.
      c. the water inside the straw will not move at all.

What Happened
What happened when you wrapped your hands around the bottle?

_____

What happened when you placed the bottle in hot water?

_____

What happened when you placed the bottle in ice water?

_____

What It Means
What do your observations tell you about how the temperature of the air in the bottle affects the movement of the water in the straw?

_____

_____

What do your observations tell you about how the temperature of the air in the bottle affects how much space the air takes up?

_____

_____

FRESHLY FROZEN

SCIENCE: Saltwater freezes at a lower temperature than plain water

STUFF: Small cup, cold tap water, ice, sealable plastic freezer bag (1-gallon size), salt, milk, sugar, vanilla, measuring cups and spoons, sealable plastic sandwich bag, gloves (optional), classroom thermometer, paper towel, spoon

What to Do

1. Fill a small cup with cold water, and add ice. Set the cup aside.
2. Fill a gallon-size plastic freezer bag about one-fourth full with ice. Add 6 tablespoons of salt.
3. Pour $\frac{1}{2}$ cup of milk, 1 tablespoon of sugar, and $\frac{1}{4}$ teaspoon of vanilla into a plastic sandwich bag. Seal the bag, and gently shake it to mix the ingredients.
4. Place the sandwich bag inside the large ice-filled bag. Seal the large bag.
5. Holding the top edge of the large bag, gently shake the bag back and forth for about 5 minutes, until the milk has turned into ice cream. You may want to wear gloves if the bag is too cold to hold.
6. Before removing the sandwich bag, take the temperature of the ice water in the large bag. Take the temperature of the ice water that you made in Step 1.
7. Take the sandwich bag out and dry the outside on a paper towel. Eat the ice cream right out of the sandwich bag. Enjoy!

What's Going On Here

The ice cream mixture changes from a liquid to a solid at 27°F (the exact freezing temperature depends on the kind of milk and the amount of sugar used). Ice water has a temperature of about 32°F, so it is not cold enough to freeze the milk and sugar. However, the salt and ice water mixture has a temperature that is below 27°F. Salt lowers the freezing temperature of water, with the exact freezing temperature of saltwater depending on the concentration of salt in the water. Since the temperature of the saltwater-ice mixture is below 27°F, the milk can freeze into ice cream. One reason you move the large bag around is to mix air bubbles inside the small bag with the milk, giving the ice cream a smoother texture. Another reason you move the bag is to expose more of the milk to the saltwater-ice mixture, making it freeze more quickly.

Try It!

Try using chocolate milk.

Try using snow instead of ice in the freezer bag. You still need to add the salt and just a little water.

FRESHLY FROZEN

What You Want to Know

How is ice cream made? How does the temperature of salt and ice water differ from the temperature of plain ice water?

What You Think Will Happen

When milk, sugar, and vanilla are mixed together and then cooled to a temperature below 27°F, they will turn into

        a. a liquid.

        b. a solid.

        c. a gas.

Plain ice water is

        a. colder than salty ice water.

        b. the same temperature as salty ice water.

        c. warmer than salty ice water.

What Happened

What happened to the milk and sugar mixture when it was shaken in a bag of ice water and salt?

_____

Record your observations in the table.

Kind of ice water	Temperature
Salty	
Plain	

What It Means

What do your observations tell you about how ice cream is made?

_____

What do your observations tell you about how salt affects the temperature of ice water?

_____

What do you think is the reason for shaking the large bag?

_____

CHAPTER 3 ELECTRICITY

Enlightening Everything Electrical

- Everything is made up of small particles called atoms. Atoms have a central part, called the nucleus, which is positively charged. Negatively charged electrons occupy the space around the nucleus of the atom. It is the electrons that play the major role in electricity.
- The buildup of electrical charge, either positive or negative, on an object is called static electricity.
- Like charges repel each other, and unlike charges attract each other. A negatively charged object will repel another negatively charged object; it will attract a positively charged object.
- You can make a neutral object appear to be positively charged by moving a negatively charged object close to it. The negatively charged object will push electrons in the other object away from that object's surface, making it appear positively charged.
- When you rub some plastic objects with wool (or felt, fur, or hair), electrons are rubbed from the wool onto the plastic. The plastic is then negatively charged, and the wool is positively charged.
- Static electricity activities work best on days that are not humid.

- The flow of electrons through a material is called electric current.
- Electrons flow easily through materials that are conductors. Electrons do not flow easily through materials that are insulators.
- As electric current flows through a material such as a wire, it encounters resistance as the moving electrons bump into atoms of the material. This resistance causes the material to heat up.
- Thin wires have more resistance to current flow than thicker wires and therefore will heat up more when a current moves through them.
- In a complete circuit, electrons can flow from the source of electric energy through objects connected to the wires in the circuit and then back to the source of energy.
- A parallel circuit is one in which the components are connected in such a way that any component makes its own complete circuit.
- In a series circuit, the components are connected to one another in such a way that the same electric current flows through all the components.

CHARGED CUPS

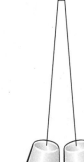

SCIENCE: Static charges can repel each other or attract each other.

STUFF: Pencil, two Styrofoam cups, string (36 inches long), two small scraps of paper, tape, piece of wool (or felt or fur)

What to Do

1. Using a sharpened pencil, poke a small hole in the center of the bottom of two Styrofoam cups.
2. Thread one end of a string through one hole and the other end through the other hole. Tie each end to a piece of scrap paper so that the string does not pull out of the cup.
3. Find the middle of the string. Tape it to the edge of a table so that the cups hang freely at about the same level. The cups will be touching unless they are already charged.
4. Rub either the inside or the outside of each cup with a piece of wool. The cups should move apart from each other.
5. While the cups are apart from each other, slowly move your hand between the cups without touching them. The cups should be attracted to your hand. If your hand doesn't move too slowly, the cups will separate again as your hand passes them.

What's Going On Here

There are two kinds of electrical charge, positive and negative. Like charges repel each other, and unlike charges attract each other. When you rub the cups with wool, they become negatively charged. Since both cups are negatively charged, they repel each other and move apart. Water molecules in the air attract electrons and slowly take away the negative charges on the cups (that's why static electricity activities work better on dry days than on humid days). As the negative charges leave the cups, they move together again because of the force of gravity. When you move your hand between the cups, the cups are attracted to your hand. This is because your hand is actually uncharged, or neutral, having an equal number of positive and negative charges. The negative charges in the cups push the negative charges in your hand away from the surface of your hand so that the surface appears positively charged. The negatively charged cups are then attracted to the positive surface of your hand.

Try It!

Try different kinds of cups.

Try different kinds of materials to charge the cups.

Try charging just one cup.

Charge a balloon by rubbing it briskly with a piece of wool. Move the balloon near an uncharged Styrofoam cup. Place it near a charged Styrofoam cup.

CHARGED CUPS

What You Want to Know

What happens to two Styrofoam cups hanging next to each other when they are rubbed with a piece of wool? What happens when you move your hand between those cups?

What You Think Will Happen

When two Styrofoam cups hanging next to each other are rubbed with a piece of wool, they will

        a. move apart from each other.
        b. move toward each other but not touch.
        c. move toward each other and touch.

When you move your hand between the cups after they have been rubbed with a piece of wool, they will

        a. move away from your hand.
        b. move toward your hand.
        c. first move toward your hand and then move away from it.
        d. first move away from your hand and then move toward it.

What Happened

What happened when you rubbed the hanging Styrofoam cups with a piece of wool?

_____

What happened when you moved your hand between the Styrofoam cups after they had been rubbed with a piece of wool?

_____

What It Means

What do your observations tell you about what kind of charges (like or unlike) are on two Styrofoam cups hanging next to each other after they have been rubbed with a piece of wool? Explain your answer.

_____

What do your observations tell you about what kind of charges (like or unlike) are on the cups and on your hands as you move your hands between the cups? Explain your answer.

_____

STATIC STICK

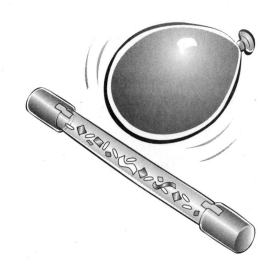

SCIENCE: Certain materials are attracted to an object with a static charge.

STUFF: Blank sheet of acetate (or overhead transparency), two plastic 35-millimeter-film canisters, tape, seven sets of small objects (for example, rice, Styrofoam bits, aluminum foil bits, scraps of paper, rice cereal), balloon, piece of wool (or felt or fur)

What to Do

1. Roll a sheet of acetate into a tube that will fit into a film canister. Put a canister on each end of the acetate to hold it in place. Tape down the edge of the acetate in a couple of spots.
2. Tape one canister in place on the end of the acetate tube.
3. Remove the canister from the other end of the tube. Drop a variety of small objects into the open end of the tube.
4. Replace the canister on the open end of the tube, and tape it in place.
5. Blow up a balloon, and tie it. Rub the balloon briskly with a piece of wool to charge it.
6. Hold the tube horizontally with the small objects spread evenly inside. Place the charged balloon close to the tube. Some of the objects should be attracted to the balloon.

What's Going On Here

When you rub the balloon with a piece of wool, electrons are rubbed from the wool onto the balloon. The balloon then has a negative charge, and the wool, having lost electrons, has a positive charge. A basic principle of static electricity states that unlike charges attract. So why does the negatively charged balloon attract the small objects in the tube? The objects in the tube are neutrally charged. When you bring the negatively charged balloon close to them, the electrons on the balloon push away the electrons on the front surfaces of the small objects so that the front surfaces of the items appear positively charged. The objects are then attracted to the balloon. (No electrons are rubbed or pushed off the small objects. The electrons are just moved away from the front surfaces. The objects actually remain neutral; it is just the front surfaces that appear positively charged.)

Try It!

Try charging the transparency tube itself, using a piece of wool, felt, or fur.

Try touching the balloon to a wall before moving it close to the tube.

Try to determine which sort of the small objects is most attracted to the tube.

STATIC STICK

What You Want to Know

What happens when you move a charged (rubbed with a piece of wool) balloon close to a clear plastic tube with small bits of various materials in it?

What You Think Will Happen

In the table below, list the materials you are going to put inside your clear plastic tube. Write an X next to the items that you think will be attracted by the charged balloon.

Material	Attracted by balloon

What Happened

In the table below, list the materials inside your clear plastic tube. Write an X next to the items that were attracted by the balloon after it was rubbed with a piece of wool.

Material	Attracted by balloon

What It Means

What do the materials that are attracted to the balloon have in common?

_____

© Dale Seymour Publications®

STICKY STATIC STUFF

SCIENCE: The force of static electricity can overcome gravity.

STUFF: Pepper, plastic wrap (12 inches x 7 inches), two straws, tape, paper towel, salt, paper clips, small scraps of paper, bits of aluminum foil, rice, small pieces of Styrofoam

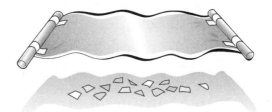

What to Do

1. Sprinkle some pepper on a table.
2. Place a piece of plastic wrap flat on the table (away from the pepper). Tape a straw along each 7-inch edge of the plastic wrap.
3. With the wrap still flat on the table, rub it briskly with the paper towel.
4. Using the straws as handles, lift the wrap off the table. Keeping the straws apart so that they are taut, slowly lower the wrap until it is about $\frac{1}{2}$ inch above the pepper. See how long the pepper will cling to the wrap.
5. Clean up the pepper.
6. Repeat Steps 1–5 for paper clips, scraps of paper, aluminum foil bits, rice, and Styrofoam bits.

What's Going On Here

There are two kinds of electrical charges, positive and negative. Positive charges push each other away, and negative charges push each other away. Positive charges draw negative charges toward them, and negative charges draw positive charges toward them. Scientists sum this up by saying that like charges repel and unlike charges attract. When you rub the plastic wrap with the paper towel, the wrap picks up electrons from the towel and becomes negatively charged. Moving the negatively charged wrap close to uncharged objects, like the pepper, causes the objects to appear positively charged. This is because the negative charges on the front surfaces of the objects race away from the negatively charged wrap. The negative charges don't actually leave the objects; they are just a little farther from the surfaces of the objects than they originally were. Since the front sides of the objects are now positive, they are attracted to the negative wrap. Several factors that will determine whether an object is attracted to the wrap are the weight and sogginess of the object, the distance the object is from the wrap, and the amount of charge on the wrap.

Try It!

Try mixing salt and pepper together, and see if you can separate the pepper from the salt using static electricity.

Try charging the wrap with different materials (for example, felt, fur, or your hair).

Measure how close you can get the wrap to the objects before they start to cling.

STICKY STATIC STUFF

What You Want to Know
What happens when you move a piece of plastic wrap that has been rubbed with a paper towel near small bits of various materials?

What You Think Will Happen
Write an X next to the items that you think will be attracted by the plastic wrap after it has been rubbed with a paper towel.

Material	Attracted by plastic wrap
Pepper	
Paper clips	
Paper	
Aluminum foil	
Rice	
Styrofoam	

What Happened
Write an X next to the items that were attracted by the plastic wrap after it had been rubbed with a paper towel.

Material	Attracted by plastic wrap
Pepper	
Paper clips	
Paper	
Aluminum foil	
Rice	
Styrofoam	

What It Means
What do materials that are attracted to plastic wrap that have been rubbed with a paper towel have in common?

_____

SHOCKS AND SPARKS

SCIENCE: An electrical charge can move through a conductor but not through an insulator.

STUFF: Plastic plate, masking tape, plastic (not paper or wax) cup (8 or 10 ounces), paper plate, aluminum foil, piece of wool, (or felt or fur)

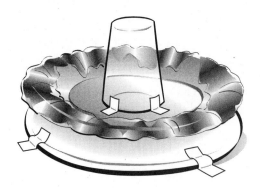

What to Do

1. Place a plastic plate on a table, facing down. Fasten the plate in place with a few pieces of tape along the edge.
2. Tape a plastic cup upside down in the middle of the top of a paper plate. The cup works as a handle. Cover the entire bottom of the paper plate with one piece of aluminum foil, wrapping the edges over the top of the plate to hold the foil in place.
3. Rub the top of the plastic plate with a piece of wool.
4. Using the cup as a handle, pick up the paper plate, and place it directly on top of the plastic plate.
5. Briefly touch your index finger to the aluminum foil–covered edge of the paper plate. You should get a small shock.
6. Using the cup as a handle, pick up the paper plate, and touch the foil-covered edge to the tip of your nose. You should get another shock.
7. Place the paper plate on the plastic plate again, and then repeat Steps 5 and 6.
8. Darken the room, and observe the sparks by repeating Steps 3–5.

What's Going On Here

When you rub the plastic plate with the piece of wool, electrons are rubbed onto the plate. The plate has a *negative* charge. Charges can build up on the plastic plate, but they cannot move around on it or off it because it is an insulator. Aluminum foil is a great conductor of electrical charges, so electrons can move freely through it. When the aluminum plate is placed on the plastic one, electrons in the aluminum foil move away from the plastic plate. That's because electrons have a negative charge and are repelled by the negatively charged plastic plate. Touching the aluminum plate with your finger gives the electrons an easy path off the aluminum plate, right to the ground. The movement of electrons off the plate is felt as a shock. In a dark room, it would show up as a small spark. Now that some electrons have moved off the aluminum plate, it is *positively* charged. Touching the plate to your nose attracts electrons from the ground through your finger back onto the aluminum plate. The aluminum plate is now uncharged, and the whole process starts over again. What you have made is a charge recycler. Touching the aluminum plate with your finger moves electrons off the plate, and touching the plate to your nose brings them back on to the plate. It may sound like the charge recycling is a property of your nose and finger. In fact, you could use your big toe and your ear. The electrons will use any path you give them to get off the plate and then back on again.

74

SHOCKS AND SPARKS

What You Want to Know

An aluminum foil–covered plate is resting on top of a plastic plate that has been rubbed with a piece of wool. What happens if you touch your finger to the edge of the aluminum-covered plate?

What You Think Will Happen

When you touch your finger to the edge of an aluminum-covered plate resting on a plastic plate rubbed with a piece of wool, you will

        a. get several small shocks.
        b. get one small shock.
        c. not get a shock at all.

When you touch the aluminum-covered plate to the tip of your nose, you will

        a. get several small shocks.
        b. get one small shock.
        c. not get a shock at all.

What Happened

What happened when you touched your finger to the edge of the aluminum-covered plate?

_____

What happened when you touched the aluminum-covered plate to the tip of your nose?

_____

What It Means

What do your observations tell you about what is needed to cause a shock?

_____

What would you change in this activity if you were going to do it again?

_____

FLUORESCENT FLASHES

SCIENCE: Static electricity can light up a fluorescent light tube.

STUFF: Balloon, piece of wool (or felt or fur), fluorescent light tube (25 watts or less), large shoe box (optional), scissors (optional), plastic plate, Styrofoam cup

What to Do

1. Darken the room.
2. Inflate a balloon, and tie it. Rub it briskly on a piece of wool.
3. Touch the balloon to the contacts on either end of a fluorescent light tube. The tube should light up momentarily. If the room is not dark enough for you to see the light, find a box large enough to hold the tube. (Eight-watt fluorescent light tubes can fit in a large shoe box.) Place the tube in the box so that contacts stick out at one end of the box. Cut out circles in the lid of the box to look into. Return to Step 2.
4. Repeat Steps 2 and 3, using a plastic plate instead of a balloon.
5. Repeat Steps 2 and 3, using a Styrofoam cup instead of a balloon.

What's Going On Here

When you rub the balloon on the piece of wool, you rub electrons off the cloth onto the balloon, and the balloon becomes negatively charged. Touching the charged balloon to the contacts of the fluorescent light tube causes that end to have a higher electric charge than the other end of the tube. A small current flows through the tube, and the bulb lights up momentarily.

Try It!

Try moving the charged balloon up and down the tube, almost touching it.

Try touching both ends of the fluorescent light tube at the same time with charged balloons.

Try different sizes and types of fluorescent lights.

Try using a regular incandescent lightbulb.

FLUORESCENT FLASHES

What You Want to Know

What happens when you touch the contacts of a fluorescent light tube with a balloon that has been rubbed with a piece of wool?

What You Think Will Happen

When you touch the contacts of a fluorescent light tube with a balloon that has been rubbed with a piece of wool, the tube will

        a. glow brightly for a few minutes.
        b. glow dimly for less than a second.
        c. glow brightly for less than a second.
        d. glow dimly for a few minutes.

What Happened

What happened when you touched the contacts of the fluorescent tube with a balloon that had been rubbed with a piece of wool?

_____

What happened when you touched the contacts of the fluorescent tube with a plastic plate that had been rubbed with a piece of wool?

_____

What happened when you touched the contacts of the fluorescent tube with a Styrofoam cup that had been rubbed with a piece of wool?

_____

Which material that you touched to the fluorescent light tube's contacts made the tube glow brightest?

_____

What It Means

What is needed to make a fluorescent light tube light up?

_____

Do you think that using a balloon that was rubbed with a piece of wool is a good way to light up a fluorescent light tube? Explain your answer.

_____

FRAZZLED FUSE

SCIENCE: When electric current flows through a very thin wire, the wire becomes hot and breaks.

STUFF: Balloon, steel wool, transparent tape, two strips of aluminum foil (1 inch x 10 inches), duct tape (or electrical tape), battery (9 volts), stopwatch

What to Do

1. Inflate a balloon, and tie it.
2. Separate out a single 3-inch-long strand of steel from a ball of steel wool.
3. Tape the halfway point of the steel strand to the center of one side of a balloon so that the strand has two loose ends.
4. Fold each of two pieces of aluminum foil in half the long way. Twist one end of one piece of aluminum foil around a loose end of the steel strand; twist one end of the other piece of foil around the other loose end of the steel strand.
5. Using electrical tape, attach the loose end of one piece of aluminum foil to one terminal of the battery. Make sure the foil touches *only one* terminal.

6. Touch the loose end of the second piece of aluminum foil to the other terminal of the battery. Oops! This step should have started out, "Plug your ears!" By now, the wire should have melted and the balloon should have popped. Note: Be careful! In this step, the battery terminal gets hot. Hold the aluminum foil so that your fingers are several inches away from the battery. Also, don't let the second piece of aluminum foil touch the battery for longer than 30 seconds. The foil too will become very hot and will drain the battery.
7. If the balloon doesn't pop within 30 seconds, try a thinner strand of steel wool or a balloon that is inflated more.

What's Going On Here

The steel wool strand heats up rapidly when you complete the circuit by touching the second piece of aluminum foil to the battery. Thin wires have a high resistance to current flow and therefore get hot when current flows through them. When the strand gets hot, it melts the balloon, and the balloon pops. The steel strand also melts in two, breaking the circuit in the process. Fuses are designed to break the circuit they are in when too much current flows through them. Breaking the circuit will keep other things in the same circuit from getting too much current and possibly becoming damaged or causing a fire.

Try It!

Try using a 1.5-volt battery.

Try the activity without the balloon. Instead, tape the steel wool strand to a piece of aluminum foil.

FRAZZLED FUSE

What You Want to Know
What happens to a very thin piece of wire when an electric current flows through it? What happens if the wire is taped to an inflated balloon?

What You Think Will Happen
When electric current flows through a very thin wire, the wire

        a. will break.
        b. will get hot.
        c. will get hot and break.
        d. will neither get hot nor break.

When electric current flows through a very thin wire taped to an inflated balloon, the balloon

        a. will pop immediately.
        b. will pop in a few seconds.
        c. will pop in a few minutes.
        d. will not pop.

What Happened
What happened to the very thin wire when you completed the circuit and electric current flowed through it?

_____

What happened to the balloon when electric current flowed through the wire?

_____

What It Means
What do your observations tell you about what happens when an electric current flows through a very thin wire?

_____

How could a very thin wire be used as a fuse, a device that cuts off the flow of electric current when the device gets too hot?

_____

_____

CONDUCTOR CHECKER

SCIENCE: Materials that are conductors can complete a circuit, causing a light to glow.

STUFF: Strip of aluminum foil (1 inch x 6 inches), string of miniature Christmas lights, scissors, two C batteries, two rubber bands, duct tape (or electrical tape), five objects to test (for example, nail, coin, eraser, pencil lead, paper clip, brass fastener, paper, toothpick)

What to Do

1. Fold a strip of aluminum foil in half the long way two times.
2. Remove from a string of miniature Christmas lights one lightbulb and about 6 inches of wire on either side. Strip about 1 inch of insulation from each end of the wire.
3. Line two batteries up so that the positive terminal of one is touching the negative terminal of the other one. Wrap two rubber bands around the batteries the long way. Wrap a piece of duct tape around the batteries the short way where the ends of the batteries meet.
4. Slide one end of the wire from the lightbulb under the rubber bands so that it touches the terminal of a battery. The rubber bands should hold it in place.
5. Attach the foil strip to the terminal of the other battery in the same way.
6. Touch the loose end of the wire from the lightbulb to the loose foil from the battery. The bulb should light up. If it doesn't, check all the connections, make sure that the batteries are touching each other and that they are not dead, and try another bulb.
7. Touch the loose end of the wire to one edge of the material to be tested, and touch the loose foil from the battery to a different point on the material. The wire should not touch the foil. Test all the materials in the same way.

What's Going On Here

Materials that conduct electricity and complete the circuit that you made are called conductors. Materials that do not conduct electricity and do not complete the circuit that you made are called insulators. The wires and foil you use in this activity are conductors.

Try It!

Make a list of conducting materials and insulating materials. Then determine which of these materials are attracted by a magnet.

Pour some water into a cup. Place the loose end of the wire and the foil in the cup, close but not touching. Mix some salt into the water. Keep adding salt until the lightbulb glows.

CONDUCTOR CHECKER

What You Want to Know

What materials allow electricity to flow through them? What materials do not allow electricity to flow through them?

What You Think Will Happen

In the table below, list the objects you are going to test. Write an X to show whether you think the object will or will not conduct electricity.

Material	Conductor	Insulator

What Happened

In the table below, list the objects you tested. Write an X to show whether the object did or did not conduct electricity.

Material	Conductor	Insulator

What It Means

Which materials that you tested conduct electricity?

_____

What do your observations tell you about the difference between materials that do and do not conduct electricity?

© Dale Seymour Publications®

ELECTRIC EXAM

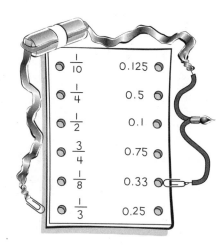

SCIENCE: When a circuit is closed, or complete, electricity will flow through it.

STUFF: String of miniature Christmas lights, two C batteries, two rubber bands, duct tape (or electrical tape), tagboard (or file folder) (7 inches x 10 inches), paper punch, marker, aluminum foil, scissors, masking tape, ruler, two paper clips

What to Do

1. Remove from a string of miniature Christmas lights one lightbulb and about 6 inches of wire on either side. Strip about 1 inch of insulation from each end of the wire.
2. Line two batteries up so that the positive terminal of one is touching the negative terminal of the other one. Wrap two rubber bands around the batteries the long way. Wrap a piece of duct tape around the batteries the short way where the ends of the batteries meet.
3. Punch six holes in each vertical (10-inch) side of a piece of tagboard.
4. List six pairs of items next to the punched holes on each side of the board so that matching items are not across from each other.
5. On the back side of the board, connect the holes for two matching items using a strip of aluminum foil that is barely wider than the punched holes. The foil should show through the punched holes. With the foil in place, carefully cover the *entire* strip with masking tape.
6. Repeat Step 5 for the other items.
7. Cut two pieces of foil 1 inch by 10 inches. Fold both of them in half the long way. Connect one piece of foil to each terminal of the battery by sliding the foil under the rubber bands.
8. Twist the loose end of one piece of foil to a wire from the lightbulb. Attach a paper clip to the end of the other wire from the lightbulb. Attach a second paper clip to the end of the remaining loose piece of foil.
9. Test your quiz board by touching one paper clip to a punched hole in one column and the other paper clip to the hole for the matching item in the other column.

What's Going On Here

The lightbulb lights up when you have a complete, or closed, circuit. This is the case when you have selected a matched pair on the quiz board. If you select the wrong match, the path for electricity is not complete, and the lightbulb will not light up.

ELECTRIC EXAM

What You Want to Know
How does a homemade electric quiz board work?

What You Think Will Happen
In a homemade electric quiz board, the lightbulb
will light up when you touch the paper clips to

        a. items that do not match.
        b. items that match.
        c. each other.

What Happened
What happened when you matched two items correctly on the quiz
board?

_____

What happened when you matched two items incorrectly on the quiz
board?

_____

What It Means
What do your observations tell you about how electricity flows in a
homemade electric quiz board circuit when you match two items
correctly?

_____

What do your observations tell you about how electricity flows in a
homemade electric quiz board circuit when you match two items
incorrectly?

_____

From what you observed about how electricity flows in a circuit, what
do you think would happen if you did not completely cover one of the
aluminum foil strips on the quiz board with masking tape?

_____

List all the materials you used to make the quiz board that are
conductors (electricity can flow through them).

_____

SERIES CIRCUIT

SCIENCE: The brightness of lightbulbs in a series circuit decreases as more bulbs are added.

STUFF: White photocopy paper, scissors, stapler, aluminum foil, ruler, masking tape, tagboard (or file folder) (9 inches x $11\frac{3}{4}$ inches), two C batteries, two rubber bands, duct tape (or electrical tape), string of miniature Christmas lights

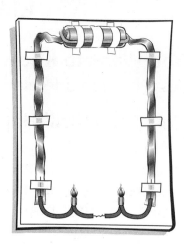

What to Do

1. Cut 10 pieces of white paper each about 2 inches square. Stack the pieces together, and staple them at one edge. This will be your brightness tester.

2. Cut two strips of aluminum foil 15 inches long and 1 inch wide. Fold each strip in half the long way. Place each strip down one of the long sides of a piece of tagboard, 1 inch in from the edge. Allow a few inches of foil to hang freely near the top of the tagboard. Use masking tape to hold the foil in place near the top and at the bottom of the tagboard.

3. Line two batteries up so that the positive terminal of one is touching the negative terminal of the other one. Wrap two rubber bands around the batteries the long way. Wrap a piece of duct tape around the batteries the short way where the ends of the batteries meet. Tape the battery set at the center near the top of the tagboard.

4. Connect one piece of loose foil to each terminal of the battery by sliding the foil under the rubber bands.

5. Remove from a string of miniature Christmas lights four lightbulbs (same color), each with about 6 inches of wire on either side. Strip about 1 inch of insulation from the ends of the wires.

6. Tape one end of a bulb's wire so that it touches one strip of foil, and tape the other end of the wire so that it touches the other strip of aluminum foil. Darken the room. Hold the brightness tester next to the bulb, and see how many pieces of paper the bulb shines through.

7. Take one end of the bulb's wire off the aluminum foil. Twist it around one end of the wire of a second bulb. Then tape the free end of the wire of the second bulb on the aluminum foil to make a complete circuit. Measure the brightness of each bulb.

8. Repeat Step 7 by adding a third bulb and then a fourth bulb.

What's Going On Here

The circuit you have made is called a series circuit. As each bulb is added to the circuit, the brightness of each bulb decreases. This is because the voltage in the circuit must be shared with each bulb.

SERIES CIRCUIT

What You Want to Know
All the circuits shown below are series circuits. What happens to the brightness of bulbs in a series circuit as more bulbs are added?

What You Think Will Happen
As more bulbs are added to a series circuit, the brightness of the bulbs
       a. will increase.
       b. will decrease.
       c. will stay the same.

What Happened
Record your observations in the blanks.

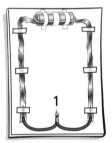

Brightness of bulb 1 _____

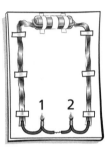

Brightness of bulb 1 _____
Brightness of bulb 2 _____

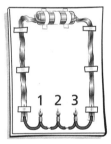

Brightness of bulb 1 _____
Brightness of bulb 2 _____
Brightness of bulb 3 _____

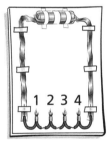

Brightness of bulb 1 _____
Brightness of bulb 2 _____
Brightness of bulb 3 _____
Brightness of bulb 4 _____

What It Means
What can you now say about how the brightness of the bulbs changes as you add more bulbs to a series circuit?

_____

PERFECTLY PARALLEL

SCIENCE: The brightness of lightbulbs in a parallel circuit stays the same as more bulbs are added.

STUFF: White photocopy paper, scissors, stapler, aluminum foil, ruler, masking tape, tagboard (or file folder) (9 inches x $11\frac{3}{4}$ inches), two C batteries, two rubber bands, duct tape (or electrical tape), string of miniature Christmas lights

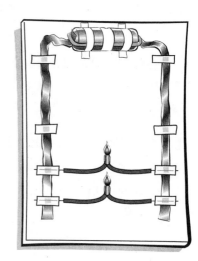

What to Do

1. Cut 10 pieces of white paper each about 2 inches square. Stack the pieces together, and staple them at one edge. This will be your brightness tester.

2. Cut two strips of aluminum foil 15 inches long and 1 inch wide. Fold each strip in half the long way. Place each strip down one of the long sides of tagboard, 1 inch in from the edge. Allow a few inches of foil to hang freely near the top of the tagboard. Use masking tape to hold the foil in place near the top and at the bottom of the tagboard.

3. Line two batteries up so that the positive terminal of one is touching the negative terminal of the other one. Wrap two rubber bands around the batteries the long way. Wrap a piece of duct tape around the batteries the short way where the ends of the batteries meet. Tape the battery set at the center near the top of the tagboard.

4. Connect one piece of loose foil to each terminal of the battery by sliding the foil under the rubber bands.

5. Remove from a string of miniature Christmas lights four lightbulbs (same color), each with about 6 inches of wire on either side. Strip about 1 inch of insulation from the ends of the wires.

6. Tape one end of a bulb's wire so that it touches one strip of foil, and tape the other end of the wire so that it touches the other strip of aluminum foil. Darken the room. Hold the brightness tester next to the bulb, and see how many pieces of paper the bulb shines through.

7. With the first bulb still in place, add a second bulb by taping one end of its wire on one strip of the aluminum foil and the other wire on the other strip of foil. Measure the brightness of each bulb.

8. Repeat Step 7 by adding a third bulb and then a fourth bulb.

What's Going On Here

The circuit you have made is called a parallel circuit. As each bulb is added to the circuit, the brightness of each bulb remains about the same. This is because the voltage in the circuit is the same across the strips of foil. Each bulb gets the same amount of voltage and thus has about the same brightness.

PERFECTLY PARALLEL

What You Want to Know
All the circuits shown below are parallel circuits. What happens to the brightness of bulbs in a parallel circuit as more bulbs are added?

What You Think Will Happen
As more bulbs are added to a parallel circuit, the brightness of the bulbs
       a. will increase.
       b. will decrease.
       c. will stay the same.

What Happened
Record your observations in the blanks.

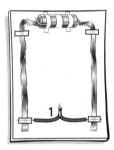

Brightness of bulb 1 _____

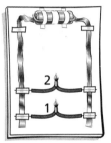

Brightness of bulb 1 _____
Brightness of bulb 2 _____

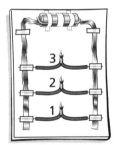

Brightness of bulb 1 _____
Brightness of bulb 2 _____
Brightness of bulb 3 _____

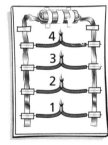

Brightness of bulb 1 _____
Brightness of bulb 2 _____
Brightness of bulb 3 _____
Brightness of bulb 4 _____

What It Means
What can you now say about how the brightness of the bulbs changes as you add more bulbs to a parallel circuit?

MARVELOUS MARKER

SCIENCE: An electric motor can be used to make a drawing machine.

STUFF: Small direct-current motor (1.5 volts), large eraser, C or D battery, rubber band, plastic cup, duct tape, marker, piece of white photocopy paper, stapler for student page

What to Do

1. Push the pin of a motor into the middle of a large eraser.
2. Wrap a rubber band around a battery the long way.
3. Set a plastic cup upside down. Tape the battery to the bottom of the cup, about in the middle.
4. Tape the motor to the bottom of the cup next to the battery. The eraser should be pointing away from the cup. It should be able to rotate freely without touching the side of the cup.

5. Tape a marker to the cup so that its point is about $\frac{1}{2}$ inch below the drinking rim of the cup.
6. Attach one of the wires from the motor to one terminal of the battery by sliding it under the rubber band. Attach the other wire in the same way to the other terminal of the battery. The eraser should start spinning.
7. Place the cup on a piece of paper, and watch it go! You can also hold the cup and make a drawing or write words.

What's Going On Here

The electric motor causes the eraser to spin. But the eraser is not perfectly balanced on the motor, so it spins unevenly and causes the cup to vibrate. When the cup is placed on the piece of paper, the vibrations are transferred to the pen, which makes squiggly marks on the paper.

Try It!

Try running the machine without the eraser on the motor.

Try different sizes of erasers.

Try taping more than one marker to the cup.

Tape the marker so that its point is farther below the rim of the cup.

MARVELOUS MARKER

What You Want to Know
What kind of machine can you make with a plastic cup, an electric motor, a rubber band, a marker, an eraser, a battery, and some duct tape?

What You Think Will Happen
What do you think a machine that you make with a plastic cup, an electric motor, a rubber band, a marker, an eraser, a battery, and some duct tape can do?

_____

What Happened
In the space below, draw what your machine looked like.

What was your machine able to do?

_____

Staple to this page an example of what your machine was able to do.

What It Means
What do your observations tell you about how to go about designing a machine?

_____

_____

How could you design another machine from the same materials? What would the new machine do?

_____

_____

CHAPTER 4 | MAGNETISM

Mystifying Magnetic Memorandum

- **A word of caution:** Keep magnets away from computers, computer disks, credit cards, videotapes, audiotapes, televisions, video recorders, tape recorders, telephones, answering machines, radios, and loudspeakers.
- Magnets have a north pole and a south pole. The north pole will point to the earth's magnetic north pole if the magnet can move freely.
- Like poles of a magnet repel each other, and unlike poles attract each other; the north pole of one magnet will repel the north pole of another magnet and attract the south pole of another magnet.
- A magnet is stronger at its poles than in between its poles.
- A magnet can exert a force on another magnet when it attracts or repels the magnet. It can also exert a force on materials that are magnetic (attracted to a magnet).
- Magnetic force can pass through some materials and can also act at some distance; a magnet can attract an object without actually touching it, even when there is some material between the magnet and the object it is attracting. Magnetic force decreases as you get farther from the magnet.

- A magnetic field exists around a magnet. Objects that are attracted to the magnet will be pulled toward the magnet more where the magnetic field is stronger.
- Objects that contain iron, nickel, or cobalt can be attracted by a magnet and can also be magnetized (made into a magnet).
- Materials that are magnetic are actually made up of small particles that are themselves tiny magnets. If these tiny magnets (also called dipoles) can be made to line up with most of their north poles facing in one direction, the material will be magnetized.
- You can magnetize some materials by placing them in a strong magnetic field.
- A wire that has electricity running through it also has a magnetic field around it.
- An electromagnet can be made by wrapping a wire around a piece of iron and then running electricity through the wire.

MIGHTY MAGNETS

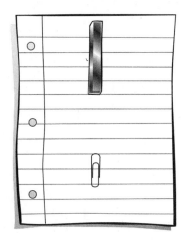

SCIENCE: The strength of a magnet is different at different points on the magnet. A magnet is strongest at its poles.

STUFF: Magnet, piece of lined notebook paper, paper clip

What to Do

1. Place a magnet on a piece of notebook paper so that the location on the magnet that you want to test is right on a line of the paper.
2. Determine how close to the magnet a paper clip can get before it is attracted to the magnet. Count the number of lines on the paper (use half lines for more accuracy) the paper clip can be from the magnet before it is attracted. That number represents the strength of the magnet in "paper line units."
3. Find the strength of the magnet in "paper line units" using another location on the magnet to attract the paper clip.

What's Going On Here

A magnet is stronger at its poles and weaker between its poles. Therefore the strength of the magnet in "paper line units" should be greatest at the poles and smallest between the poles. A bar magnet is usually shaped like a rectangle with the poles located on the short sides.

Try It!

Try testing the relative strength of different magnets.

Try testing the strength of two magnets together.

Try placing something (for example, paper, a book, or pencils) between the magnet and the paper clip and then testing the strength of the magnet.

MIGHTY MAGNETS

What You Want to Know

How can you test the strength of a magnet? Where on a magnet is it strongest?

What You Think Will Happen

Draw a picture of the magnet that you are going to use. On the magnet, place an S where you think that it will be strongest and a W where you think it will be weakest.

What Happened

Record your observations in the table.

Part of magnet tested	Strength (paper line units)

Draw a picture of the magnet that you used. On the magnet, place an S where it was strongest (had the greatest number of "paper line units"). Place a W where it was weakest (had the smallest number of "paper line units").

What It Means

A magnet is strongest at its poles. Where are the poles of your magnet located?

_____

MORE MIGHTY MAGNETS

SCIENCE: The strength of a magnet can be determined in several ways.

STUFF: Small magnets, paper clips, masking tape

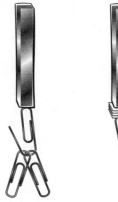

Method 1   Method 2

What to Do

Use the following two methods to determine the strength of a magnet:

Method 1

1. Bend the edge of a paper clip out a bit so that other paper clips can hook onto it. Using a piece of rolled masking tape, attach a small magnet to the side of a table.

2. Place the bent paper clip at the location on the magnet that you want to test. Hang one paper clip at a time on the bent one, and count the number of paper clips that the magnet can hold before they all drop. This is the strength of the magnet measured in "paper clip units."

Method 2

1. Put a piece of masking tape at the location on the magnet that you are testing, and try to pick up a paper clip with the magnet.

2. Continue adding pieces of tape, one at a time, until the magnet does not pick up the paper clip. (You may want to add five pieces of tape to the magnet at a time in the beginning if you have a strong magnet.) The number of pieces of tape on the magnet is a measure of its strength in "tape units."

Test the strength of another location on the same magnet using the same two methods.

What's Going On Here

The strength of a magnet can be determined in many different ways. One way is to determine how much weight it can hold; this is done using the paper clips. The more paper clips that the magnet can support at a point, the stronger the magnet is at that point. Another way of measuring the strength of a magnet is to measure how much of a material, like tape, its force can pass through. The more tape the magnetic force has to pass through, the weaker it becomes, until finally it can no longer attract a magnetic object like a paper clip. The stronger the magnetic force of a magnet, the more pieces of tape it can pass through and still support the paper clip. If a magnet (or location on a magnet) is weaker than another magnet (or location on a magnet) using one method, it should be weaker for another method as well.

MORE MIGHTY MAGNETS

What You Want to Know
How can you test the strength of a magnet?

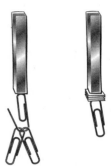

What You Think Will Happen
Draw a picture of the magnet that you are going to use. On the magnet, place an X where you think that it will be strongest.

What Happened
Draw a picture of the magnet that you used. On the magnet, place a P where it was strongest using method 1 (paper clips) and a T where it was strongest using method 2 (layers of tape).

What It Means
What can you now say about how different methods of testing a magnet's strength compare with each other?

Which was your favorite way of testing a magnet's strength? Why?

GIVING GRAVITY GRIEF

SCIENCE: The force of magnetism can compete with the force of gravity and win!

STUFF: Masking tape, strong magnet, paper clip, string, 10 index cards, aluminum foil

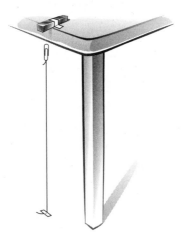

What to Do

1. Using masking tape, secure the magnet to the edge of a table. Leave part of the magnet exposed. Or make things really simple, and attach the magnet to a metal cabinet or a refrigerator. Note: Don't attach it to anything it could damage!
2. Tie a paper clip to the end of a piece of string.
3. Put the paper clip on the exposed part of the magnet.
4. Slowly, using the string, pull the paper clip away from the magnet in a downward direction. If you are very careful, you will be able to suspend the paper clip in the air a short distance from the magnet.
5. Tape the other end of the string onto something nearby so that the paper clip is suspended in air a short distance from the magnet.
6. Carefully try to place an index card between the magnet and the paper clip. Keep increasing the number of cards, one at a time, until the paper clip drops.
7. Repeat Step 6, using small pieces of aluminum foil.

What's Going On Here

The paper clip is suspended in air because the magnetic force of the magnet extends beyond the magnet itself. Gravity is a force that is pulling the paper clip down, but the force of magnetism is pulling the paper clip up. The force of gravity and the magnetic force have a tug-of-war over the paper clip, and the stronger force will win. When the paper clip is too far from the magnet or when too many index cards (or pieces of aluminum foil) are placed between it and the magnet, gravity wins and the paper clip drops. When the paper clip is close enough to the magnet, the magnetic force wins and the paper clip is suspended in the air. The magnetic force acts through various materials besides air, as you show by placing paper or aluminum foil between the magnet and the paper clip.

Try It!

Try placing a butter knife between the magnet and the paper clip.

Try using other magnets.

Try using something besides the paper clip to suspend midair.

Try placing other materials between the magnet and the paper clip.

Try pulling the paper clip away from the magnet horizontally.

GIVING GRAVITY GRIEF

What You Want to Know

What happens when you hold a paper clip attached to a string close to a magnet? Will a magnetic force go through paper? Will a magnetic force go through aluminum foil?

What You Think Will Happen

When you move a paper clip attached to a string close to a magnet, the magnet will attract (pull toward itself) the paper clip

       a. only if they touch.

       b. if they're close but not touching.

       c. even if they're several feet apart.

How many index cards do you think you can place between a paper clip and a magnet before the magnet stops attracting the paper clip?

_____

How many pieces of aluminum foil do you think you can place between a paper clip and a magnet before the magnet stops attracting the clip?

_____

What Happened

What happened when you moved the paper clip attached to the string close to the magnet?

_____

In the table below, record the number of pieces of a material that you placed between the paper clip and the magnet before the magnet could no longer attract the paper clip.

Material	Number of pieces before paper clip fell
Index cards	
Aluminum foil	

What It Means

What do your observations tell you about how a magnetic force can act through air, paper, and aluminum foil?

_____

MAGNETIC MAPPINGS

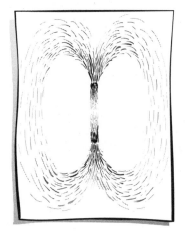

SCIENCE: Magnets have a magnetic force field around them that can be modeled with magnetic field lines.

STUFF: Magnet, piece of white photocopy paper, iron filings, old saltshaker, spray adhesive (or hair spray)

What to Do
Note: Iron filings can be obtained from several sources, including school supply stores, science supply catalogs, and children's magnetic toys. If you are patient, you can also find them at the beach or in some sandboxes. Put a magnet in a plastic sandwich bag, and drag the bag through the sand. Brush off into a container the iron bits that the magnet picks up. Put the filings in a saltshaker with big enough holes that you can shake the filings out.

1. Place a magnet flat on a table.
2. Lay a piece of paper over the magnet.
3. Gently sprinkle iron filings on the paper lying on the magnet. Make sure you sprinkle the filings over the magnet and several inches away from the magnet too. Tap the paper gently to spread the filings out evenly.
4. While the paper is still lying on the magnet, spray the paper with spray adhesive. Wait a few minutes until the spray is dry, and lift the paper from the magnet.

What's Going On Here
Magnets always have a magnetic field around them. You can't see the field, but you can visualize it by observing its effect on iron filings. The iron filings help to map out the magnetic field because they fall right along the magnetic field lines and tend to be concentrated where the field is strongest and more sparsely scattered where the field is weaker.

Try It!

Try making a pattern using two bar magnets with like poles facing each other about 1 inch apart, and then again with unlike poles facing each other at that distance.

Try using two disk-shaped magnets close to each other.

Try other orientations of different kinds of magnets.

MAGNETIC MAPPINGS

What You Want to Know
What happens when you sprinkle iron filings on a piece of paper that is covering a magnet? Where will most of the iron filings settle?

What You Think Will Happen
Draw the shape of the magnet you are going to use. Then draw the pattern that you think the iron filings will settle in when you sprinkle them on a piece of paper covering the magnet.

What Happened
Draw the shape of the magnet you used. Then draw the pattern that the iron filings settled in when you sprinkled them on the piece of paper covering the magnet.

What It Means
Most of the iron filings will settle around the area on the magnet that is strongest. What do your observations tell you about where the strongest part of your magnet is located?

_____

What do your observations tell you about where the weakest part of the magnet is located?

_____

MAGNETIC MONEY

SCIENCE: A magnet can attract some forms of money and not others.

STUFF: Strong magnet, penny, nickel, dime, quarter, dollar bill

What to Do

1. Try to attract coins—a penny, nickel, dime, and quarter, in turn—with a magnet. Touch the magnet to each coin, trying to pick it up.
2. Try to attract a dollar bill with the magnet in the same way that you did with the coins.

3. Fold the dollar bill in thirds, and place it on a table with its middle third flat and its ends pointing upward.
4. Move the magnet very close to but not quite touching the printed border of one of the ends that is sticking up.

What's Going On Here

When you try to attract coins with a magnet, the force of gravity is stronger than any force of attraction between the magnet and any magnetic materials in the coins. Depending on the strength of the magnet, you probably are not able to lift the dollar bill up with the magnet either.

However, when you position the dollar bill in such a way that you do not have to contend with the force of gravity so much, you are able to attract the bill slightly. The inks used to print the dollar bill have magnetic material in them and so are attracted by the magnet.

Try other U.S. currency, such as a five-dollar bill.

Try foreign coins and currency.

Try to balance a U.S. coin on its edge and make it roll by attracting it with a magnet.

Try hanging a dollar bill by a string and attracting it with a magnet.

MAGNETIC MONEY

What You Want to Know
Which U.S. coins and paper money are attracted (pulled toward) a magnet? What is the best way to position money so that it is easily attracted by the magnet?

What You Think Will Happen
U.S. money that will be attracted by a magnet includes the
a. penny.
b. nickel.
c. dime.
d. quarter.
e. dollar bill.

What Happened
Which U.S. coins and paper money were attracted by the magnet?

_____

How did you have to position the money so that the magnet would attract it?

_____

What It Means
What do your observations tell you about which U.S. coins and paper money are attracted by a magnet?

_____

What do your observations tell you about the best way to position money so that a magnet can easily attract it?

_____

SEPARATING SAND

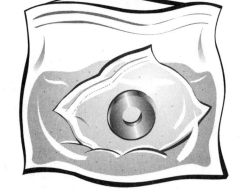

SCIENCE: A magnet can be used to sort materials.

STUFF: Sealable plastic sandwich bags, paper clips, brass fasteners, pennies, salt, iron filings, playground sand, strong magnet, paper

What to Do

Note: Iron filings can be obtained from several sources, including school supply stores, science supply catalogs, and children's magnetic toys. If you are patient, you can also find them at the beach or in some sandboxes. Put a magnet in a plastic sandwich bag, and drag the bag through the sand. Brush off into a container the iron bits that the magnet picks up.

1. In one sandwich bag, put a mixture of paper clips, brass fasteners, and pennies.
2. In a second bag, put a mixture of salt and iron filings.
3. In a third bag, put about a cup of playground sand.
4. Put a strong magnet inside an empty plastic bag, and seal the top. Place the sealed magnet into the first bag, and shake it gently.
5. Remove the magnet bag, and gently shake it to remove any nonmagnetic particles. Carefully remove the magnet from its bag so that the magnetic particles fall onto a piece of paper.
6. Repeat Steps 4 and 5 with the other two bags, making sure to remove the magnetic material from the magnet bag each time.

What's Going On Here

The objects that are magnetic will stick to the magnet even through the plastic bag that surrounds the magnet. This shows that magnetic force can go through materials. Magnets are often used in industry to sort materials. You may have heard of cow magnets, very strong cylindrical magnets that are placed in a cow's stomach. The cow magnet stays in the cow's stomach its entire life collecting bits of wire and other metals that the cow eats so that they don't injure the cow internally.

Try It!

Try sorting out magnetic material from a sample of soil.

Find the mass of the magnetic material in each bag and the mass of the nonmagnetic material in each bag. Determine the ratio of magnetic to nonmagnetic material.

Make up your own combinations of materials to place in a sealable plastic sandwich bag for sorting.

SEPARATING SAND

What You Want to Know
How can a magnet be used to separate one material from another?

What You Think Will Happen
A first plastic bag contains paper clips, brass fasteners, and pennies. What will a magnet be able to pull out of this bag?

_____

A second plastic bag contains a mixture of salt and iron filings. What will a magnet be able to pull out of this bag?

_____

A third plastic bag contains sand. What will a magnet be able to pull out of this bag?

_____

What Happened
Record your observations in the table.

Bag	Materials pulled out by magnet
1	
2	
3	

What It Means
What do your observations tell you about how a magnet can be used to separate materials?

_____

_____

What do your observations tell you about what might happen if you put a magnet into a bag of *soil* from the playground?

_____

CEREAL SLURRY

SCIENCE: The iron powder in some high-iron cereals can be separated from the cereal using a magnet.

STUFF: Measuring cup, water, blender, high-iron cereal (such as Total cereal), clear plastic cup (8 or 10 ounces), strong magnet, plastic spoon, two other kinds of cereal

What to Do

1. Pour $1\frac{1}{2}$ cups of water into a blender. Add 1 cup of dry high-iron cereal, and blend at high speed for about 30 seconds. You should have a smooth slurry.
2. Fill a plastic cup about three-fourths full of the slurry.
3. Place a strong magnet against the outside of the cup about halfway up from the bottom.
4. While holding the magnet against the cup, stir the slurry with the plastic spoon for about 30 seconds.
5. Pull the magnet away from the cup, and look at the spot on the cup where the magnet was touching.
6. Move the magnet around while holding it against the cup.
7. Repeat Steps 1–5 for each of the other cereals. Note which of the three cereals attracts the most particles to the magnet.

What's Going On Here

Many iron-rich cereals contain iron in a powdered form that is baked into the cereal. When you blend the high-iron cereal with water, the iron powder separates from the rest of the cereal and is suspended in the slurry (insoluble matter mixed in water). When you hold the magnet against the side of the cup, the magnetic force goes through the plastic and attracts the iron powder to it. The iron powder remains stuck to the cup for a short time after you pull the magnet away. You can make the powder follow the magnet around the cup by moving the magnet. Depending on the amount and form of iron that is in the other cereals, you may or may not be able to see any powder on the side of the cup. It is interesting that the same metal that is used to make automobiles, hammers, and bicycles is used in a powder form in cereals!

Try It!

Try other kinds of cereal.

Try other kinds of food.

Try using cereal that has been blended without using water.

Seal the magnet inside a plastic sandwich bag, and place it in the slurry. Move the bag around the slurry for a few minutes, and then remove it.

CEREAL SLURRY

What You Want to Know
Is the iron in iron-rich cereals attracted to a magnet?

What You Think Will Happen
In the table below, list the cereals you are going to use. Write an X next to the ones you think will have particles that are attracted to a magnet.

Cereal	Particles attracted to magnet

What Happened
Which cereal had the most particles attracted to the magnet?

_____

For that cereal, what did you observe when you pulled the magnet away from the cup after you had been holding it there for a while?

_____

What did you observe when you moved the magnet around the cup?

_____

What It Means
Look at the list of ingredients, vitamins, and minerals on each box of cereal. Which item do you think is attracted to a magnet?

_____

Look at the nutritional facts on each box of cereal. Which cereal has the highest daily-requirement amount of the item that you think is attracted to a magnet?

_____

Do your observations about which cereal was most attracted agree with the numbers on the cereal boxes? Explain.

_____

MAGICAL MAGNET

SCIENCE: A needle can be made into a magnet. A current-carrying wire has a magnetic field around it.

STUFF: Sewing needle, magnet, small piece of paper, dish of water, pocket compass (optional), flexible insulated wire (8 inches long), scissors, rubber band, C or D battery, glove

What to Do

1. Rub a sewing needle the long way about 50 times on a magnet in the same direction. Lift the needle after each stroke.
2. Place the needle on a small scrap of paper, and float it in the middle of a dish of water. Make sure the needle is balanced so that one end doesn't dip into the water. Also make sure that the magnet you used to magnetize the needle is very far away!
3. The north pole of the magnet-needle should point north. Check this by using a compass to find north. Move the magnet-needle so that it doesn't point north, and watch it swing back north again.
4. Strip about 1 inch of insulation from both ends of a piece of insulated wire. Bend the wire so that it has a U-shape with a flat bottom that is about 2 inches long. Hold the wire close to but not touching the floating magnet-needle; be

sure the flat bottom of the wire is close to the magnet-needle. The wire should have no effect on the magnet-needle.
5. Put a rubber band around the battery the long way. Attach the wire ends to the battery terminals underneath the rubber band.
6. Hold the battery itself as you move the flat bottom of the wire close to but not touching the floating magnet-needle. You should be unable to get the magnet-needle to line up with the wire. Note: The current in the wire will quickly drain the battery, and the ends of the wire will become hot. Do not leave the wire connected to the battery for more than a minute, and don't touch the exposed ends of the wire. Put a glove on before removing the wire from the battery terminals. The battery will also become warm, so be careful handling it.

What's Going On Here

Rubbing the needle repeatedly on the magnet causes the iron atoms in the needle to align in such a way that the needle becomes a temporary magnet with a north pole and a south pole. Putting the needle-magnet in the water causes it to float with its north pole pointing north. A current-carrying wire also has a magnetic field associated with it. The direction of the magnetic field can be determined using the right-hand rule: Pretend that you are grasping the wire with your right hand and pointing your thumb in the direction the current moves as it goes from the positive terminal to the negative terminal of the battery. Your fingers will point in the direction of the magnetic field. The magnetic field is thus perpendicular to the wire, and when the wire is held close to the magnet-needle, the magnet-needle lines up with the wire's magnetic field, perpendicular to the wire.

MAGICAL MAGNET

What You Want to Know

What happens when you hold a wire with a current near a needle that has been magnetized?

What You Think Will Happen

After you magnetize a needle by rubbing it repeatedly on a magnet, you float it on a scrap of paper in a dish of water. When you move the bottom part of a U-shaped piece of wire near the floating needle,

     a. the needle will move so that it is lined up with the wire.
     b. the needle will move so that it makes a plus sign with the wire.
     c. the needle will not move at all.

When you move the bottom part of a U-shaped piece of wire that has an electric current in it near the floating needle,

     a. the needle will move so that it is lined up with the wire.
     b. the needle will move so that it makes a plus sign with the wire.
     c. the needle will not move at all.

What Happened

What happened when you placed the bottom part of a U-shaped piece of wire near the floating needle?

_____

_____

What happened when you placed the bottom part of a U-shaped piece of wire that has an electric current in it near the floating needle?

_____

_____

What It Means

What can you now say about what happens when a wire carrying a current is placed near a floating magnetized needle?

_____

_____

You have made two magnets. What are they, and how did you make them?

_____

_____

ENERGETIC ELECTROMAGNET

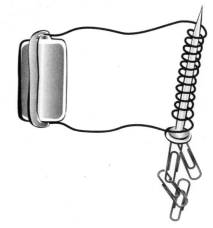

SCIENCE: A current-carrying wire wrapped around a steel nail can make the nail into a temporary magnet.

STUFF: Steel nail, small paper clips, flexible insulated wire (about 2 feet long), scissors, rubber band, C or D battery

What to Do

1. Touch the head of a nail to a pile of paper clips to see if it will pick them up.
2. Wrap a piece of wire tightly around the nail in coils, leaving about 6 inches of wire free at both ends. Strip about 1 inch of insulation from both ends of the wire.
3. Touch the head of the nail to the paper clips to see if it will pick them up.
4. Wrap a rubber band around a battery the long way. Slide one end of the wire under the rubber band so it touches one terminal of the battery. Slide the other end of the wire under the rubber band so it touches the other terminal of the battery.

5. Immediately try to pick up the paper clips with the head of the nail. Note: The current in the wire will quickly drain the battery, and the ends of the wire will become hot. Don't leave the wire connected to the battery for longer than a minute, and don't touch the exposed ends of the wire. The battery will also become warm, so be careful handling it.
6. Carefully remove the wires from the battery. Try to pick up the paper clips with the head of the nail.
7. Remove the coil of wire, and try to pick up the paper clips with the head of the nail.

What's Going On Here

When a wire is hooked up to a battery, a current flows in the wire because you have completed the circuit. A current-carrying wire has a magnetic field surrounding it. The magnetic field surrounding the wire induces magnetism in the nail, making it behave like a magnet. The nail is thus able to attract paper clips. When the current is cut off from the wire, the nail is still able to pick up paper clips because the magnetic field from the current-carrying wire has turned the nail into a temporary magnet. The nail remains magnetized for a while even after the magnetic field from the current-carrying wire is removed.

Try It!

Try picking up paper clips with a coil of current-carrying wire with no nail inside it.

Try wrapping a coil of wire around a paper clip to determine if it will become magnetized.

Try picking up large paper clips.

ENERGETIC ELECTROMAGNET

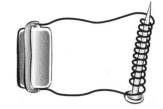

What You Want to Know

How can you make a magnet using a coil of wire wrapped around a steel nail?

What You Think Will Happen

For each step in the table below, predict whether the head of a nail will pick up paper clips.

Step	Will nail pick up clips?
1. The nail alone is used.	
2. A wire is wrapped around nail.	
3. Each end of the wire is attached to an end of the battery.	
4. The ends of the wire are removed from the battery.	
5. The wire is removed from the nail.	

What Happened

For each step in the table below, record your observations.

Step	Will nail pick up clips?
1. The nail alone is used.	
2. A wire is wrapped around nail.	
3. Each end of the wire is attached to an end of the battery.	
4. The ends of the wire are removed from the battery.	
5. The wire is removed from the nail.	

What It Means

The magnet you've made is called an electromagnet. What do your observations tell you about how an electromagnet is made?

CHAPTER 5 ▌SOUND

Some Splendid Sound Stuff

- Sound is caused by vibrations—that is, something moving back and forth.
- Most sounds we hear travel through the air, but sound can travel through any material.
- Whenever an object vibrates in any material, sound is produced.
- Frequency is the number of times in a second that a vibrating object moves back and forth. The frequency of the vibrating object is also the frequency of the sound produced.
- Pitch is the highness or lowness of the sound and is related to the frequency of the sound.
- The faster an object vibrates, the higher the frequency and pitch of the sound it makes. Likewise, the slower an object vibrates, the lower the frequency and pitch of the sound.
- In general, sound travels faster through liquids (like water) than it does through gases (like air). Sound travels faster through solids (like wood) than it does through liquids. The speed of sound in air is about one-fifth of a mile per second, or about 1 mile every 5 seconds.
- As you get farther away from the source of a sound, the intensity, or loudness, of the sound decreases.
- Sound can be reflected off objects. If the object that sound is reflecting off is far enough away, the reflected sound may be heard as an echo.

SPRINGING SPRINKLES

Hum or sing into here.

SCIENCE: Sound is caused by vibrations.

STUFF: Oatmeal or salt box, ruler, knife, wax paper, rubber band, candy decorating sprinkles (or salt), mirror (optional)

What to Do

1. If you are using a salt box, cut off the end with the spout. If you are using an oatmeal box, take off the lid.
2. Cut a $1\frac{1}{2}$-inch-square hole in the side of the box about an inch from the bottom.
3. Place wax paper over the open end of the box, and secure it with a rubber band. Pull the wax paper so that it is tight on the open end of the box.
4. Place the box on a table. Sprinkle cake decoration sprinkles on the wax paper.

5. Hum or sing into the square hole in the box, and observe the movement of the sprinkles. You may need a partner to watch for you, or you can hold the box in front of a mirror while you hum or sing into it.
6. Try humming or singing notes that are higher and lower in pitch to see how the sprinkles vibrate. Try humming or singing louder and softer.
7. Try blowing air into the hole in the box to see if the sprinkles will vibrate.

What's Going On Here

Sound is caused by vibrating air. When you hum or sing into the box, you are causing the air inside the box to vibrate, or move back and forth. The vibrating air causes the wax paper to vibrate, and the vibrating paper causes the sprinkles on top to move about. Vibrations have been transferred from your vocal cords to the air to the wax paper and finally to the sprinkles. Higher notes should cause the paper to vibrate

more rapidly and the sprinkles to move about more rapidly. You may be able to hit a certain note at which the sprinkles jump noticeably higher from the wax paper. This note corresponds to the natural frequency of the oatmeal box; the sound waves are just the right length so that, when they vibrate the wax paper, they reinforce each other, causing the sprinkles to jump higher.

Try It!

Try using different materials—such as aluminum foil, fabric, or regular paper— to cover the top of the oatmeal box.

Try speaking at various volumes, and notice differences in how the sprinkles respond.

Try playing some music from a tape player into the oatmeal box.

Try covering the hole with wax paper and humming or singing near it.

SPRINGING SPRINKLES

What You Want to Know

What happens when you hum or sing into a box that has one end covered by wax paper with candy sprinkles on top?

What You Think Will Happen

When you hum into a box that has one end covered by wax paper with candy sprinkles on top, the candy sprinkles will

       a. pop up and down on the wax paper.

       b. move side to side but not leave the wax paper.

       c. not move much at all.

       d. jump right off the box.

If you hum into the box more loudly,

       a. the candy sprinkles will move less.

       b. the candy sprinkles will move more.

       c. the candy sprinkles will move the same as before.

When you blow air into the box, the candy sprinkles will

       a. pop up and down on the wax paper.

       b. move side to side but not leave the wax paper.

       c. not move much at all.

       d. jump right off the box.

What Happened

What happened when you hummed into the box that has one end covered by wax paper with candy sprinkles on top?

_____

What happened when you changed how loudly you hummed?

_____

What happened when you blew air into the box?

_____

What It Means

What do your observations tell you about what makes sound and what sound can do?

_____

BUSY BUZZER

SCIENCE: Sound is produced by a vibrating object.

STUFF: Styrofoam ball (1-inch diameter), scissors, index card (3 inches x 5 inches), craft stick (or tongue depressor) ($\frac{3}{4}$ inch x 6 inches), stapler, string (about 30 inches), rubber band ($\frac{1}{8}$ inch thick, 6-inch perimeter)

What to Do

1. Cut a Styrofoam ball in half, using the sharp edge of a pair of scissors.
2. Line up the long edge of an index card with the long edge of a craft stick so that about $\frac{1}{2}$ inch of the stick shows at each end.
3. Staple the index card to the stick in three places.
4. Push the Styrofoam hemispheres onto each end of the stick so that their flat edges are facing each other.

5. Tie one end of a piece of string around one end of the stick between the index card and Styrofoam hemisphere. Grab the free end of the string, and twirl the "buzzer" around your head.
6. Stretch a rubber band around the two hemispheres. Straighten it so that it does not have any twists.
7. Grab the free end of the string, and twirl the buzzer around your head.
8. Try changing the speed that you twirl the buzzer.

What's Going On Here

As you twirl the "buzzer" around your head, you should hear a humming sound similar to that of a large mosquito or bee. The index card keeps the buzzer level as you twirl it through the air, and the air causes the rubber band to vibrate. The vibrating rubber band produces the humming sound that you hear.

Try It!

Try making a "buzzer" without the index card.

Try a rubber band with a different thickness on your buzzer.

Try a rubber band with a different tightness on your buzzer.

Try different sizes and shapes of card on your buzzer.

Try making a large buzzer with a paint-stirring stick, sheet of poster paper, and large rubber band.

BUSY BUZZER

What You Want to Know
How is sound made by a homemade "buzzer" that you can twirl around your head?

What You Think Will Happen
In a homemade "buzzer," the thing that will vibrate (move back and forth) to make the sound you hear is the
> a. index card.
> b. cut Styrofoam balls.
> c. string.
> d. rubber band.

When you twirl a homemade buzzer faster,
> a. the pitch will become higher.
> b. the pitch will become lower.
> c. the sound will become louder.
> d. the sound will become softer.

What Happened
What happened when you twirled the homemade buzzer around your head *without* the rubber band attached?

_____

What happened when you twirled the homemade buzzer around your head *with* the rubber band attached?

_____

What happened when you changed the speed that you twirled the buzzer?

_____

What It Means
List all the things that vibrate in a homemade buzzer. Which of these things make the buzzing sound?

_____

What can you now say about how the speed at which the rubber band vibrates changes the pitch of the sound you hear from a buzzer?

_____

WHIRLING WOODEN WONDER

SCIENCE: Vibrating air produces sound.

STUFF: Paint-stirring stick (or ruler), string (about 5 feet long), scissors with pointed tip

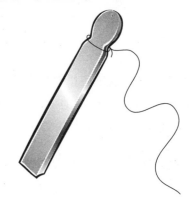

What to Do

1. Tie a string around the notched end of a paint stirrer (or into a hole on a ruler).
2. Go outside where you will have a lot of room away from other people, trees, and buildings.
3. Holding the free end of the string, twirl the stick in a circle over your head. Twirl the string faster and faster, and notice how the sound changes.

4. If you can't get the stick to make a roaring sound, twirl it vertically, allowing it to hit the ground a few times to get the stick wiggling, and then twirl it over your head.
5. If you still can't get the stick to make a roaring sound, poke a hole in the middle of the top part of the stick using the pointed end of a pair of scissors. Tie the string there. Repeat Step 3.

What's Going On Here

As the stick is twirled through the air, the stick vibrates to produce the roaring sound that you hear. As the stick spins through the air, it winds up the string in one direction and begins to spin lower toward the ground. Then it unwinds the string and starts to rise as it is spinning. The pitch of the sound the spinning stick makes changes as it winds and unwinds the string. Changing the speed of the twirling stick changes the pitch and the volume of the sound. The twirling stick is similar to the aboriginal musical instrument called a bullroarer. The bullroarer is also used as a means of communication among the Aborigines.

Try It!

Try using a stick with a different width.

Try using a stick with a different length.

Try using a different length of string.

Try twirling the stick in a vertical circle.

See how far a partner can get from you and still hear the roaring sound.

With a partner, try to work out a code, using the twirling wooden stick to communicate.

WHIRLING WOODEN WONDER

What You Want to Know
What happens when you twirl a wooden stick attached to a string in a circle over your head?

What You Think Will Happen
When you twirl a wooden stick on a string over your head, the thing that will vibrate (move back and forth) to make the sound you hear is
> a. the string.
> b. the stick.

When you twirl the stick faster,
> a. the pitch will become higher.
> b. the pitch will become lower.
> c. the sound will become louder.
> d. the sound will become softer.

What Happened
What happened when you twirled the wooden stick over your head?

_____

What happened when you changed the speed at which you twirled the wooden stick?

_____

What kind of sound did the string make?

_____

What It Means
What do your observations tell you about how sound is produced?

_____

What do your observations tell you about how the speed at which a wooden stick is twirled changes the pitch of the sound you hear from the stick?

_____

CRAZY KAZOO

Hum into here.

SCIENCE: Sound is caused by vibrations.

STUFF: Toilet paper tube, wax paper, rubber band, paper punch

What to Do

1. Hum into one of the open ends of a toilet paper tube.
2. Wrap a piece of wax paper around one of the open ends of the tube. Place a rubber band around the wax paper to hold it in place. Pull the wax paper so that it is tight on the end of the tube. Hum into the open end of the tube.
3. Place one of your fingers gently on the wax paper as you hum into the tube. You should be able to feel the vibrations.
4. Use a paper punch to punch a hole near the open end of the tube. Hum into the open end of the tube.
5. Repeat Step 3.
6. Hum a song into the tube.

What's Going On Here

Sound is caused by vibrations. As you hum into the tube, vibrations from your vocal cords cause the air to vibrate. The vibrating air causes the wax paper to vibrate, and that produces the kazoo noise that accompanies your humming.

Try It!

Try punching a hole in the end of the kazoo near the wax paper.

Try other lengths of tube (for example, a paper towel or a wrapping-paper tube).

On the end of the tube, instead of wax paper try other materials—for example, aluminum foil, paper, or fabric.

CRAZY KAZOO

What You Want to Know
How is sound made by a homemade kazoo?

What You Think Will Happen
When you hum into either end of an open tube, the resulting sound will
> a. be the same as your voice.
> b. be louder than your voice.
> c. have a buzzing quality.
> d. be softer than your voice.

When you hum into the open end of a tube with wax paper covering one end, the resulting sound will
> a. be the same as your voice.
> b. be louder than your voice.
> c. have a buzzing quality.
> d. be softer than your voice.

When you hum into the open end of a tube with wax paper on one end and a hole near the open end, the resulting sound will
> a. be the same as your voice.
> b. be louder than your voice.
> c. have a buzzing quality.
> d. be softer than your voice.

What Happened
Describe the sound when you hummed into the empty tube.

_____

Describe the sound when you hummed into the tube with wax paper covering one end.

_____

Describe the sound when you hummed into the tube with wax paper covering one end and a hole punched near the other end.

_____

What It Means
What can you now say about how sound is made in a kazoo?

_____

INTRIGUING INSTRUMENT

SCIENCE: Sound is caused by vibrations. Sound travels better through solids (like yarn) than through gases (like air).

STUFF: Yarn (36 inches long), string (36 inches long)

What to Do

1. Place your hands flat over your ears.
2. Have a partner drape the middle part of a piece of yarn around the back of your head and over your hands so that the loose ends are in front of you. Use your thumbs to hold the yarn in place.
3. Have the partner grab the loose ends of the yarn with one hand, pull them taut, and strum them. You should hear pleasant musical tones.

4. Have your partner shorten the length of the yarn by moving her or his hands up and then strum again.
5. Trade places with your partner, and repeat Steps 1–4.
6. Try using string, and see what effect that has on the pitch of the tones.

What's Going On Here

Sound is caused by vibrations. When your partner strums the yarn, it starts to vibrate. These vibrations start the air around the yarn vibrating, and your partner is able to hear the sound from the yarn. You are also able to hear the sound of the strummed yarn. Vibrations from the yarn travel directly to your hands, which then transmit the vibrations to your ears. You are able to hear the sound from the vibrating yarn much better than your partner for two reasons. First of all, vibrations from the yarn that travel through the air travel in all directions, and your partner picks up only the tiny fraction of available waves that hits her ears. You, on the other hand, pick up most of the vibrations traveling directly through the string to your hand. The second reason the sound is louder for you is that sound travels better through solids (like yarn) than through gases (like air). Sound waves must be passed from one molecule to the next through the medium that the sound waves are traveling through. The molecules are farther apart in a gas than they are in a solid, so sound travels better through a solid.

Try It!

Try yarn having a different thickness or length.

Try playing a simple song by varying the tension in (tightness of) the yarn.

Try placing small paper cups instead of your hands over your ears. Put notches in the rims on the cups' bottoms to hold the yarn in place. Pull the yarn to the front, and hold it taut. This way you can play the instrument by yourself.

INTRIGUING INSTRUMENT

What You Want to Know

How can sound be made from a piece of yarn that is wrapped around your head and over your ears?

What You Think Will Happen

When someone strums a piece of yarn that is wrapped around your head and over your ears, the sound will be

         a. louder for you.

         b. louder for the person doing the strumming.

         c. about the same loudness for both of you.

When the piece of yarn is shortened and someone strums it,

         a. the pitch of the sound will be higher than before.

         b. the pitch of the sound will be lower than before.

         c. the sound will be louder than before.

         d. the sound will be softer than before.

What Happened

Describe the sound that you heard when your partner strummed the piece of yarn that was held over your ears.

_____

Describe the sound that you heard when you did the strumming with the yarn on your partner's ears.

_____

What happened to the sound you heard from the yarn over your ears when the length of the yarn was shortened?

_____

What It Means

What do your observations tell you about the difference between how sound travels through solids (like yarn) and through gases (like air)?

_____

What do your observations tell you about how the sound changes when you shorten the length of a piece of yarn being strummed?

_____

SILLY SEAL SOUNDS

SCIENCE: Sound is produced by vibrations.

STUFF: Paper cup, pencil, kite string (24 inches long), small piece of sponge (1 inch x $1\frac{1}{2}$ inches), water, toothpick

What to Do

1. Punch a hole in the middle of the bottom of a cup, using a sharpened pencil.
2. Tie one end of a piece of string around the center of the sponge.
3. Dampen the sponge with water. Fold the sponge around the string at the loose end, and pull the sponge down the length of the string. Listen to the sound.
4. Push the loose end of the string through the hole in the bottom of the cup from the inside. Pull the loose string all the way to the outside. Tie a toothpick to this loose end of the string.
5. Then pull the sponge end of the string taut so that the toothpick is tight against the bottom of the cup.
6. Fold the sponge around the string about at the drinking rim of the cup, and pull the sponge down the length of the string. Listen to the sound.

What's Going On Here

When you pull the dampened sponge along the string in Step 3, you hear a small sound. But when you pull the dampened sponge along the string attached to the cup, the cup produces a rather loud sound similar to that of a barking seal. Sound is produced by vibrations. As you pull the dampened sponge down the length of string, it actually moves in a jerking motion due to friction between the sponge and the string. The jerking motion of the sponge along the string causes the string to vibrate. The vibrating string causes the cup to vibrate, producing the sound that is picked up by your ears. The sound is loud because the cup acts like a megaphone, concentrating and directing the sound waves.

Try It!

Try using cups made of different materials (for example, plastic and Styrofoam).

Try pulling the sponge down the string in short jerks.

Try different sizes of cups.

Try different types of string.

Try placing your thumb and index finger around the string. Pull downward along the string with dry fingers. Then moisten your fingers in water, and try again.

SILLY SEAL SOUNDS

What You Want to Know

What kind of sound can you make by rubbing a wet sponge along a string? What will happen if you do the same thing when the string is attached to a paper cup?

What You Think Will Happen

When you pull a dampened sponge along a string,

        a. you will hear a soft sound.

        b. you will hear a loud sound.

        c. you will hear no sound.

When you pull a dampened sponge along a string that is attached to a paper cup,

        a. you will hear a soft sound.

        b. you will hear a loud sound.

        c. you will hear no sound.

What Happened

What happened when you pulled the dampened sponge along the string when it was not attached to the cup?

_____

What happened when you pulled the dampened sponge along the string when it was attached to the cup?

_____

What did the noise sound like?

_____

What It Means

What do your observations tell you about how you can make a sound using a string and a sponge?

_____

What do your observations tell you about what you can do to change the loudness of a sound?

_____

METALLIC MUSIC

SCIENCE: Sound travels better through solids (like string) than it does through gases (like air).

STUFF: Metal spoon, string (about 3 feet long), metal hanger, metal cooling rack for baking

What to Do
1. Tie the handle of a metal spoon in the middle of a piece of string.
2. Wrap the ends of the string around your index fingers. (Or make a loop in both ends of the string to put your index fingers through.)

3. Gently swing the spoon so that it taps the edge of a table. Listen to the sound.
4. Repeat Step 3, but this time put your fingers in your ears immediately after the spoon taps the table.
5. Repeat Steps 1–4 for a metal hanger and a metal cooling rack.

What's Going On Here
Sound is produced when an object vibrates. When the spoon taps the table, the spoon starts to vibrate. The string attached to the spoon then vibrates, followed by your finger, your eardrum, and the small bones in your ear. You hear the spoon vibrations much better when you have your fingers in your ears for two reasons: First, when the spoon vibrates, it causes the air to vibrate, producing sound waves that are picked up by your ear. The sound waves produced by the vibrating spoon travel in all directions, and your ears pick up only a fraction of the available waves. When you put your fingers in your ears, the sound is more pronounced because the vibrations travel directly along the string to your ear. The second reason the sound is louder when your fingers are in your ears is that sound travels better through solids (like string) than through gases (like air). Sound waves must be passed from one molecule to the next in the medium that the sound waves are traveling through. In air, the molecules are farther apart, so sound does not travel so well as through a solid, where the molecules are much closer together.

Try It!

Try different lengths of string or a string with a different thickness.

Try listening with just one ear.

Instead of wrapping the free ends of the string around your index fingers, attach them to two cups. Hold the cups over your ears.

Try attaching two strings to the ends of a metal or plastic Slinky toy.

Try a plastic hanger or one that is coated with vinyl.

METALLIC MUSIC

What You Want to Know
What is the difference between how sound travels through solids (like string) and how it travels through gases (like air)?

What You Think Will Happen
A metal object is tied to a string that is wrapped around your index fingers. When the object taps the edge of a table, the sound will be

        a. louder when you listen to the sound without putting your fingers in your ears.

        b. louder when you immediately put your fingers in your ears.

        c. the same either way.

What Happened
Describe the sound that you heard when the metal spoon tapped the table and you listened without your fingers in your ears.

_____

Describe the sound that you heard when the metal spoon tapped the table and you immediately put your fingers into your ears to listen.

_____

Describe the sounds you heard using the metal hanger and the metal cooling rack.

_____

What It Means
What do your observations tell you about the difference between how sound travels through solids (like string) and through gases (like air)?

_____

What could you do to make the sound from the metal spoon louder?

_____

What could you do to change the pitch of the sound from the metal spoon?

_____

SENSATIONAL SOUNDS

SCIENCE: Pitch changes as an object vibrates at different rates. The faster an object vibrates, the higher the pitch of the sound.

STUFF: Ruler

What to Do

1. With one hand, hold a ruler against a tabletop so that most of the ruler hangs off the edge of the table. Be sure that your hand holding the ruler is directly over the *edge* of the table so that just the end of the ruler that is hanging off the table can vibrate.
2. With the other hand, strum the end of the ruler that is hanging over the edge of the table.
3. Move the ruler so that less of it is hanging off the edge of the table, and strum it again. The pitch of the sound should be higher.
4. Continue to shorten the length of the ruler hanging off the edge of the table.
5. Place your ear against the tabletop as you strum the ruler. The sound should be louder than when you heard the sound through air.

What's Going On Here

Sound is produced by vibrating objects. In this activity, the vibrating ruler causes the air to vibrate; that in turn causes your eardrum to vibrate. The shorter length of ruler vibrates faster and thus produces a higher pitch than the longer piece. When you place your ear on the tabletop, the sound is louder than when you hear it through air because sound travels better through solids (like wood) than it does through gases (like air).

Try It!

Try a wood or plastic ruler of a different length or thickness.

Try to play a simple tune by yourself on a ruler.

Try to play a simple tune with several others, each of you playing one note of the tune on a ruler.

Try using a comb or another object.

SENSATIONAL SOUNDS

What You Want to Know
What happens when you strum the end of a ruler that is hanging off the edge of a table?

What You Think Will Happen
When you allow more of a ruler to hang off the edge of a table, the sound you hear when you strum it will be

        a. louder.
        b. softer.
        c. higher in pitch.
        d. lower in pitch.

When you place your ear on a tabletop as you strum a ruler, the sound you hear will be

        a. louder.
        b. softer.
        c. higher in pitch.
        d. lower in pitch.

What Happened
What happened when you changed the length of the ruler hanging off the edge of the table?

_____

How did the sound change when you placed your ear on the tabletop as you strummed the ruler?

_____

What It Means
What can you now say about how the length of a ruler that is vibrating (moving back and forth) affects the sound that you hear?

_____

What can you now say about how the sound from a vibrating ruler changes when you listen to it through a solid, like a tabletop, rather than through a gas, like air?

_____

List all the things that vibrate when you strum a ruler.

_____

STRAW SOUNDS

SCIENCE: The pitch of the sound you make as you blow across the top of a straw changes as the length of the column of air in the straw changes.

STUFF: Drinking glass, water, straws, ruler, scissors, index card (3 inches x 5 inches), tape

Blow here. →

What to Do

1. Fill a glass with water.

2. Put a straw into the water. Place your lower lip against the top of the straw, and blow over the top of the straw to make a sound.

3. Slowly pull the straw out of the water while you are blowing over the top of it. Then blow over the straw as you slowly lower it back into the water.

4. Cut straws to the lengths of 1 inch, 3 inches, 5 inches, and 7 inches. Tape the straws in order of length on an index card, allowing 1 inch between them. The straws should be level on top and be about $\frac{1}{2}$ inch above the top of the index card.

5. Place your lower lip against the top of each straw and blow over the top of the straw to make a sound.

What's Going On Here

As you blow over the top of the straw, you are causing the air inside the straw to vibrate, producing the sound you hear. As the length of the vibrating column of air gets shorter, the air vibrates faster and the pitch increases. You have a higher pitch with a shorter straw (or a straw that is well into the water) than you have with a longer straw (or a straw that is well out of the water).

Try It!

Try adding more straws to the index card.

Try playing a tune on the straws attached to the index card.

Try putting clay into the bottom of the straws on the index card to determine if that affects the sound.

STRAW SOUNDS

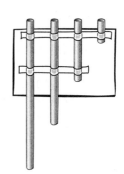

What You Want to Know

What happens when you blow over the top of a straw as you lower the straw into water? What happens when you blow over the top of straws of different lengths?

What You Think Will Happen

When you blow over the top of a straw as you lower the straw into water, the sound you hear will

a. become louder.

b. become softer.

c. become higher in pitch.

d. become lower in pitch.

When you blow over the top of a short straw, the sound will be

a. louder than when you blow over a long straw.

b. softer than when you blow over a long straw.

c. higher in pitch than when you blow over a long straw.

d. lower in pitch than when you blow over a long straw.

What Happened

Describe how the sound changed when you blew over the top of a straw as you pulled the straw out of a glass of water.

_____

Describe how the sound changed when you blew over the top of a straw as you lowered the straw into a glass of water.

_____

Describe how the sound changed as you blew over the tops of straws of different lengths.

_____

What It Means

Trumpets and trombones use vibrating air tubes to make music. What do your observations tell you about what must change for these instruments to make the various pitches of a tune? How do you think you make that change?

_____

BALLOON BOX BLASTER

SCIENCE: Sound is caused by vibrations. Vibrating air can also make things move and can put out a candle.

STUFF: Oatmeal or salt box, knife, pencil, large balloon, scissors, short candle, matches

What to Do

1. If you are using a salt box, cut off the end with the spout. If you are using an oatmeal box, take the lid off. Using a sharpened pencil, punch a hole in the middle of the bottom of the oatmeal or salt box between $\frac{1}{4}$ and $\frac{1}{2}$ inch in diameter.

2. Cut off the top of a balloon a third of the way down from the blowing end. Take the blowing end of the balloon and blow into it (this isn't really part of the activity, but it's fun to do and it shows that vibrations produce sound). Stretch the rest of the balloon over the open end of the box so that the end looks like a drum.

3. Light the candle.

4. Lay the box on its side. Be sure the hole is at about the same height as the candle flame. Point the hole in the box toward the flame, and position the box about 6 inches from the flame.

5. Hold the box steady with one hand. Pull back on the center of the balloon with the other hand, and let go of the balloon. The balloon should bounce back toward the box, and the flame should go out.

What's Going On Here

When you pull back on the balloon and then release it, the balloon vibrates, causing the air inside the box to vibrate. The vibration is passed on from the air particles inside the box to those outside the box and eventually to the air particles right in front of the flame. The flame goes out when the moving air near the flame vibrates over it.

Try It!

Try using an empty soup can with a hole punched in the bottom.

Try using a large coffee can.

See how far the flame can be from the box or can and still be put out.

Try using a smaller or larger hole punched in the box or can. You can do this without having to get a new can or box each time. Just punch a hole in the box or can that is the largest you want to try. Then tape a piece of aluminum foil over the large hole, and punch the smaller hole in the aluminum foil. Change the aluminum foil when you want to change the size of hole.

BALLOON BOX BLASTER

What You Want to Know
A balloon is stretched over the open end of a round box. The other end of the box has a small opening in it. A candle is placed close to the end of the box with the hole in it. What happens when you pull back on the balloon and let it go?

What You Think Will Happen
When you pull back on the balloon and let it go, the candle
     a. will glow brighter.
     b. will flicker but not go out.
     c. will go out.
     d. _____.

What Happened
What happened to the candle when you pulled back on the balloon and let it go?

_____

Describe the sound that you heard when you pulled back on the balloon and let it go.

_____

What It Means
What can you now say about how air can put a candle out?

_____

List all the things that vibrate (move back and forth) when you pull back on a balloon on a homemade balloon blowing box?

_____

HANDY HEAR-O-METER

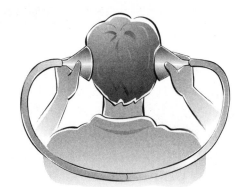

SCIENCE: The volume of sound decreases as it travels farther.

STUFF: Permanent marker, plastic tubing (36 inches long) to fit funnels (available at hardware stores), two small funnels, butter knife, cotton ball

What to Do

1. Mark a line on a length of plastic tubing to divide it into two equal parts.
2. Place a funnel on each end of the tubing.
3. Sit on a chair with your back to a table. Place the funnels over your ears and rest the plastic tubing on the table behind you.
4. Have a partner tap the tubing with a butter knife while the tubing is resting on the table. The person should tap the tubing about 8 inches from the midpoint. Raise your hand to show in which ear the tapping sound is louder.
5. Have your partner continue to tap on opposite sides of the midpoint mark on the tube, and see how close he or she can get to the mark before you can't distinguish in which ear the sound is louder.
6. Place a cotton ball tightly in the right funnel opening. Repeat Steps 3–5.

What's Going On Here

The vibrations made by tapping the knife travel through the tube toward your ears. The volume of the sound decreases as it travels a greater distance. If the point where you can't distinguish the louder sound is at the midpoint of the plastic tubing, your ears hear about equally well. If the point is closer to your right ear, it means that your left ear hears better. If the point is closer to your left ear, it means that hearing in your right ear is better. This test does not tell you whether you have a hearing problem in either ear; it tells you only which ear has the *better* hearing.

Try It!

Try putting a piece of cloth over one of your ears.

Using one funnel, see how far away from the funnel your partner can tap before you can't hear the sound anymore. Try this with both ears to see which ear hears better. Do the results of this test agree with the results of the test using both ears at the same time?

HANDY HEAR-O-METER

What You Want to Know
What happens to the loudness of sound as you get farther from the source of the sound?

What You Think Will Happen
Write an X on the drawing below to show the location of tapping that you think you will be able to hear with about the same loudness in each ear. Write a C on the drawing for the location of tapping that you think you will be able to hear with about the same loudness in each ear when a cotton ball is placed in the right funnel.

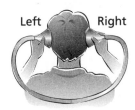

Left   Right

Middle of tube

What Happened
Write an X on the drawing below to show the location of tapping that you were able to hear with about the same loudness in each ear. Write a C on the drawing for the location of tapping that you were able to hear with about the same loudness in each ear when a cotton ball was placed in the right funnel.

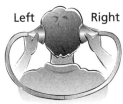

Left   Right

Middle of tube

What It Means
What do your observations tell you about the loudness of sound as you get farther from the source of the sound?

_____

If a person can hear tapping equally well in both ears when the tapping is closer to the left ear than to the right, what can you say about that person's hearing?

CHAPTER 6 | LIGHT

Luminous Light Lore

- Light has wave properties. The wavelength of a light wave is the distance between peaks in the wave, and the frequency is the number of waves that pass by per second.

- The color of light can be determined by either its wavelength or its frequency. Red light has a long wavelength (low frequency), and violet light has a short wavelength (high frequency). The brightness of light is called its intensity.

- White light is made up of all the colors of the rainbow (or spectrum): red, orange, yellow, green, blue, indigo, and violet. A prism can be used to disperse white light into its seven component colors.

- When light strikes an object, it can be reflected (bounced back from the surface), transmitted (passed through the object), or absorbed (neither transmitted nor reflected).

- Most objects are seen by reflected light. Objects generally reflect various amounts of all the wavelengths of light, and the eye combines them to give the sensation of the color that you see.

- When light bounces off materials, it obeys the law of reflection, which says that the angle at which light strikes a surface is the same as the angle at which it is reflected from the surface.

- The farther you get from a source of light, the dimmer the light becomes.
- The speed of light in air is slightly less than it is in a vacuum, about 186,000 miles per second. Light travels about a million times faster than sound.
- The speed of light is slower in air than in a vacuum and slower still in glass and plastics. The change in the speed of light as it travels from one material into another causes the light to bend, or refract. Lenses refract light in very useful and interesting ways.
- When two or more light waves come together, they sometimes interfere with one another and combine to create a wave with properties (wavelength, frequency, and intensity) that may be different from any of the original waves.
- Light is diffracted when it hits an obstacle. Diffraction effects are most noticeable when the object is very small, like a speck of dust on your eyeglasses.

FAST FLICKER FLASHES

SCIENCE: Since your eye can hold an image for only a fraction of a second, presenting two images to the eye very quickly causes the two to merge into one.

STUFF: Two unlined index cards (3 inches x 5 inches), pencil with eraser, ruler, crayons (or markers), duct tape, stapler

What to Do

1. On each index card, use a pencil to lightly draw a 2-by-2-inch square in the middle of the card.
2. In the square on one of the index cards, draw a goldfish bowl, using the whole area inside the square.
3. In the square on the other index card, draw a goldfish slightly smaller than the square and right in the middle. Color both pictures.
4. Erase the squares you drew on each index card.
5. Use duct tape to attach one card to the pencil. The pencil should be centered on the back of the card.

6. Place the other card so that it is back-to-back with the first card and exactly lined up with it. Check that both the pictures are right side up, and staple the cards together at their corners.
7. Hold the pencil upright between the palms of your hands, and twirl the pencil so that the pictures on the two cards alternate quickly in front of your eyes. You should see the fish inside the fishbowl. Vary the speed at which you twirl the pencil.

What's Going On Here

The retina of the human eye can hold onto an image for a fraction of a second. If another image hits the retina of the eye before the first one has faded away, the two merge and are seen as one image. If you move the pictures of the fishbowl and the fish fast enough, they merge into one image of the fish inside the bowl.

 **Try It!**

Try making other picture combinations. A child and a swing, a bird and a cage, a dog and a doghouse, and a tree and apples are examples of drawings to try.

Try making word combinations—for example, S I N E S U ! on one card and C E C I F N on the other card. Be careful to line the letters up so that in this example when you spin the cards, you read: SCIENCE IS FUN!

Try the activity in a dark room with a flashlight aimed at the card.

FAST FLICKER FLASHES

What You Want to Know

What happens when you flash two different pictures in front of your eyes very rapidly?

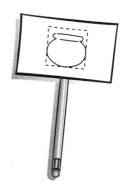

What You Think Will Happen

A card has a picture of a fish. On the back is a card that has a picture of a fishbowl. If you flash the two cards rapidly in front of your eyes, you will see

        a. just the fish.
        b. just the fishbowl.
        c. the fish inside the fishbowl.
        d. a blur.

What Happened

Draw a picture of what you saw when you flashed the pictures from the two cards in front of your eyes.

How did what you saw change when you changed how fast you twirled the pencil?

_____

What It Means

What do your observations tell you about what you see when two different pictures are flashed in front of you very rapidly?

_____

What do your observations tell you about what happens when you flash the pictures too quickly?

_____

What do your observations tell you about what happens when you flash the pictures too slowly?

_____

SHARP SHADOWS

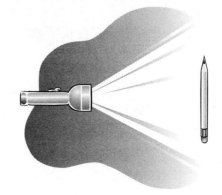

SCIENCE: Shadows vary in their sharpness, depending on the source of light.

STUFF: Pencil, flashlight, wax paper, rubber band

What to Do
1. Hold a pencil about a foot from a wall in a darkened room.
2. Hold a flashlight about a foot away from the pencil, and shine it toward the pencil so that a shadow appears on the wall.

3. Cover the front part of the flashlight with wax paper, using a rubber band to hold the wax paper in place. Repeat Steps 1 and 2. Observe the shadow.

What's Going On Here
When you shine the flashlight on the pencil, the pencil blocks some of the light from the flashlight and a shadow appears on the wall. When light shines directly from the flashlight, it is confined to a relatively narrow beam so that it shines directly at the wall. When you put wax paper over the front of the flashlight, the wax paper scatters the light from the flashlight and the beam is no longer narrow and directed. The shadow produced from the wax-paper-covered flashlight is fuzzy. The edges of the shadow are not as sharp as they were when you didn't use the wax paper; the shadow isn't as dark either because some of the light from the flashlight is absorbed and scattered by the wax paper. Similarly, your shadow outside on an overcast day is fuzzier than your shadow on a clear day. You will also notice that when the wax paper is over the flashlight, details of the flashlight bulb are not as noticeable around the shadow of the pencil as they are when you use the flashlight directly. Again, the wax paper has scattered light in many directions, so details of the flashlight bulb are lost.

Try It!

Try observing your shadow outside on a sunny day and on an overcast day.

Try varying the distance of the pencil from the wall when you are making the shadow.

Try varying the distance of the flashlight from the pencil when you are making the shadow.

Try placing several layers of wax paper over the flashlight.

Try placing fabric over the flashlight.

SHARP SHADOWS

What You Want to Know

What is the shadow of a pencil like when you use a flashlight as the source of light? How is the shadow of the pencil different when you cover the front part of the flashlight with wax paper?

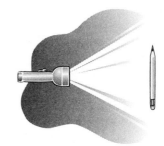

What You Think Will Happen

The shadow of a pencil made using a flashlight will be

        a. fuzzy and dark.

        b. fuzzy and not too dark.

        c. sharp and dark.

        d. sharp and not too dark.

The shadow of a pencil made using a flashlight with the front covered with wax paper will be

        a. fuzzy and dark.

        b. fuzzy and not too dark.

        c. sharp and dark.

        d. sharp and not too dark.

What Happened

Describe the shadow of the pencil when you used the flashlight.

_____

Describe the shadow of the pencil when you used the flashlight covered with wax paper.

_____

Which shadow showed the most details of the flashlight bulb and reflector?

_____

What It Means

What can you now say about the difference between shadows made using only a flashlight and using a flashlight covered with wax paper?

_____

PINHOLE PICTURES

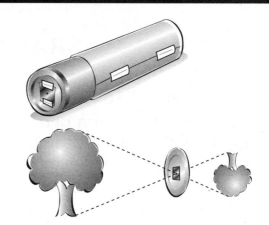

SCIENCE: The image in a pinhole camera is smaller than the object and is upside down.

STUFF: Small can (10 ounces), hammer, nail, aluminum foil, tape, sewing pin, wax paper, rubber band, black construction paper (12 inches x 9 inches)

What to Do

1. Carefully clean the inside of an empty can. Using a hammer and a nail, punch a hole in the middle of the bottom of the can.
2. Tape a piece of aluminum foil over the hole in the bottom of the can.
3. Make a hole in the aluminum foil with a sewing pin. Move the pin around a little to widen the hole. The pinhole should go through the hole you punched in Step 2 but should not be as wide.
4. Put a piece of wax paper over the open end of the can. Pull it tight, and secure it with a rubber band.
5. Wrap the 9-inch side of a piece of construction paper around the can so that it covers about 2 inches of the end of the can with the wax paper and extends beyond that end of the can. Tape the paper in place.
6. Staying indoors, hold the construction paper close to one eye as you point the pinhole end of the can outside at a tree on a sunny day. Look at the image of the tree on the wax paper.

What's Going On Here

The pinhole camera you made has an upside-down image projected on the wax paper. The image is smaller than the actual object. The drawing above demonstrates why the image is upside-down and smaller. Many rays of light leave an object. We can consider just a couple of them to show where the image of the object is located and what its orientation will be. The construction paper serves as a light shield to keep room light from hitting the wax paper and reducing the contrast of the image.

Try It!

Try looking at a bright object (for example, a lamp or a candle) indoors.

Try changing the size of the pinhole.

Try changing the size of the can.

Staying indoors, try looking at an outdoor object using the camera without the light shield.

Try going outside with the camera and looking at various objects.

PINHOLE PICTURES

What You Want to Know

A pinhole camera consists of a can with a small hole punched in one end and a piece of wax paper covering the other end. What kind of picture will you see on the wax paper when you point the end with the pinhole at a bright object?

What You Think Will Happen

The picture of an object on the wax paper of a pinhole camera will be

        a. upside down and smaller than the object.
        b. upside down and larger than the object.
        c. right side up and smaller than the object.
        d. right side up and larger than the object.

What Happened

Describe the picture that you saw in your pinhole camera.

_____

_____

How did the black construction paper help you to see the picture in the pinhole camera?

_____

_____

What It Means

What do your observations tell you about the kind of picture that is made in a pinhole camera?

_____

_____

A pinhole camera is similar to the retina of your eye, the back part of the eye where pictures of what you look at are formed. What kind of picture do you think appears on the retina of your eye?

_____

_____

How is the pinhole camera similar to a real camera? How is it different?

_____

_____

_____

FEARLESS FLAME

SCIENCE A piece of glass reflects and transmits light.

STUFF: Clay, piece of glass (from picture frame), clear plastic cup (8 or 10 ounces) (slightly larger than candle), ruler, matches, small candle, water

What to Do

1. Using clay for support, stand a piece of glass vertically on a table.
2. Place an empty clear cup so that the middle of the bottom of the cup is 8 inches from the middle of one side of the piece of glass.
3. Using clay for support, stand a candle on the other side of the piece of glass plate about 8 inches from the middle of the glass.

4. Light the candle, and darken the room.
5. Look at the piece of glass from the candle side of the glass. You should be able to see the candle and the empty cup. It should look like the candle is positioned in the cup.
6. Slowly pour water into the cup. Look at the piece of glass from the candle side. It should look like the candle is burning under water.

What's Going On Here

The piece of glass both reflects and transmits light; it reflects about 4% of the light that strikes it and transmits about 96%. When you look at the glass from the candle side, you see the cup because 96% of the light from the cup is transmitted through the glass. You also see the reflec- tion of the candle because 4% of the light coming from the candle is reflected by the glass. The image of the candle is located as far behind the glass as the candle is in front of the glass. So the reflection of the candle is 8 inches behind the glass, right in the middle of the cup!

Try It!

Try looking at the glass plate from the cup side of the glass.

Try changing the positions of the cup and the candle.

Try using other objects on either side of the glass.

FEARLESS FLAME

What You Want to Know

What happens when you look into a glass plate with a lighted candle on your side and an empty clear cup on the other side?

What You Think Will Happen

When you look at a glass plate that has a lighted candle on your side and a clear plastic cup on the other side, you will

        a. not see the candle.
        b. not see the cup.
        c. see neither the candle nor the cup.
        d. see both the candle and the cup.

What Happened

What did you see when you looked at the glass plate with the lighted candle on your side and the clear plastic cup on the other side?

_____

_____

What happened when water was poured into the clear plastic cup?

_____

What It Means

What do your observations tell you about how light is transmitted by (passed through) a glass plate?

_____

_____

What do your observations tell you about how light is reflected (bounced back) by a glass plate?

_____

_____

REAL REFLECTIONS

SCIENCE: The angle at which light hits a mirror is the same as the angle at which it bounces off the mirror.

STUFF: Pencil, ruler, piece of white photocopy paper, protractor, clay, small mirror, scissors, black construction paper, tape, flashlight

What to Do

1. Draw a line down the middle of a piece of paper; call this line A. Place the flat edge of a protractor (corresponding to 0° and 180°) on this line. The angle measure of 90° should be marked 0°. Mark the angles on either side of 90° in steps of 10°. The angle measures of 100° and 80° should be marked 10°; angle measures of 110° and 70° should be marked 20°; and so on. Draw lines from line A to each of the marked angles.

2. Use clay to prop up a mirror so that it stands vertically with one edge along the line on the paper and so that the shiny side of the mirror faces the angle lines.

3. Cut a circle out of a piece of black construction paper that will fit over the lighted end of a flashlight. Cut a vertical slit about $\frac{1}{8}$ inch wide in the middle of

the construction paper circle. The length of the slit should be slightly less than the diameter of the circle. Tape the circle on the flashlight.

4. Set the flashlight on the piece of white paper 2 inches away from the mirror. Be sure the slit in the construction paper circle is vertical.

5. Turn the flashlight on, and shine it toward the mirror. Then darken the room. Move the flashlight until the beam of light strikes the mirror at an angle of 10°. The beam should strike the mirror at the point where all the lines meet. Notice at what angle the beam is reflected.

6. Repeat Step 5 for all the other angles you have marked.

What's Going On Here

The angle at which light hits a mirror is the same as the angle at which it bounces off the mirror. In other words, the angle of incidence (incoming angle) is equal to the angle of reflection (outgoing angle). This is called the law of reflection. The angles are usually measured from a perpendicular line that is drawn to the point on the mirror where the light strikes it. That is why you marked the angles from 10° to 90° on either side of the perpendicular.

 **Try It!**

Try propping up a second mirror parallel to the first one but a few inches away so that light will bounce off the first mirror and hit the second mirror. Does the law of reflection still work?

Try using a sheet of white paper taped to the mirror.

REAL REFLECTIONS

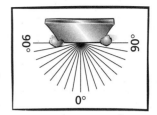

What You Want to Know

How is the angle at which light is reflected from a mirror related to the angle at which it hits the mirror?

What You Think Will Happen

The angle at which light is reflected from a mirror will be

       a. the same as the angle at which it hits the mirror.

       b. smaller than the angle at which it hits the mirror.

       c. greater than the angle at which it hits the mirror.

What Happened

Record your observations in the table.

Angle at which light hit mirror	Angle at which light was reflected from mirror
0°	
10°	
20°	
30°	
40°	
50°	
60°	
70°	
80°	
90°	

What It Means

What can you now say about how the angle at which light is reflected from a mirror is related to the angle at which light hits the mirror?

_____

What happens to a beam of light that hits a mirror straight on (0° in this activity)?

_____

MIRRORS MULTIPLYING MONEY

SCIENCE: Mirrors that are placed at an angle to each other produce multiple reflections.

STUFF: Two mirrors (1 inch square or larger), duct tape (or electrical tape), protractor, penny

What to Do

1. Tape two mirrors together along one edge, putting the tape on the unshiny side of the mirrors.
2. Stand the mirrors upright so that the angle between the shiny sides is 120°.
3. Place a penny between the two mirrors.
4. Look into the mirrors, and count the total number of pennies you see, including the real penny.
5. Repeat Steps 2, 3, and 4 for angles of 90°, 60°, and 45°.

What's Going On Here

When a penny is placed in front of a mirror, there is an image of the penny in the mirror. If you count the image of the penny and the penny itself, there are two pennies. If you hinge two mirrors together and place them at right angles to each other, you will count four pennies, the penny itself and three images in the two mirrors.

The added images are due to multiple reflections between the two mirrors or, in other words, reflections of reflections. The number of pennies that you will see (images plus the real penny) can be determined by dividing 360° by the number of degrees between the two mirrors.

Try It!

Try other angles between the mirrors.

Try adding a third mirror by taping it to the edge of one of the first two mirrors.

Try other objects instead of a penny.

Try angles that are not evenly divided into 360°.

Try placing the penny closer to or farther away from the mirror to see what effect distance has on the image.

Take the tape off the mirrors, and use clay to stand them upright so that their shiny sides are facing each other about 4 inches apart. Place the penny between the two mirrors, and try to count the images.

MIRRORS MULTIPLYING MONEY

What You Want to Know
You see just one reflection in one mirror. How many reflections do you see in two mirrors that are touching at an angle?

What You Think Will Happen
When the two mirrors touch at a 120° angle and a penny is between them, you will see (counting the real penny and any reflections)

        a. 1 penny.
        b. 3 pennies.
        c. 120 pennies.
        d. too many pennies to count.

When two mirrors touch at a 90° angle and a penny is between them, you will see (counting the real penny and any reflections)

        a. 1 penny.
        b. 4 pennies.
        c. 90 pennies.
        d. too many pennies to count.

What Happened
Record your observations in the table.

Angle between mirrors	Number of pennies (real penny plus reflections)
120°	
90°	
60°	
45°	

What It Means
What do your observations tell you about the number of pennies you see as you make the angle between two mirrors smaller?

_____

What angle should the two mirrors be placed at for you to see seven pennies? What information did you use to make this prediction?

_____

RADICAL REFLECTIONS

 SCIENCE: Multiple images are produced when light from an object reflects off several mirrors.

STUFF: Three pieces of Plexiglas (1 inch x 6 inches), two pieces of duct tape (each about 2 inches wide and 6 inches long), blank sheet of acetate (or overhead transparency), scissors, ruler, permanent markers in several colors, nail

What to Do

1. Look at your reflection in a piece of Plexiglas.
2. Lay one piece of duct tape on a table, sticky side up.
3. Place one piece of Plexiglas flat in the middle of the tape, going lengthwise. Look at your reflection in the Plexiglas.
4. Place the second and third pieces of Plexiglas with their long edges touching the long edges of the first piece. Bend the second and third pieces inward so that their upper long edges touch each other. Pull the excess tape along the edges of the first piece up so that the tape holds the second and third pieces of Plexiglas in place. The short sides of the three pieces of Plexiglas should form a triangle.
5. Lay the long side of a second piece of duct tape along the long side of the second piece of Plexiglas, just overlapping the first piece of duct tape. Fold the duct tape around the uncovered pieces of Plexiglas so that the triangular tube is completely covered with duct tape.
6. Hold the tube close to but not touching your eye, and look at objects around the room.
7. Cut two circles ($2\frac{1}{4}$ inches in diameter) out of the acetate. Make designs on both of them using the permanent markers. Make squiggles, lines, letters, and so on, but try not to make the design solid. Be creative!
8. Use a nail to punch a hole in the center of the circles, and push the circles to the head of a nail. Tape the nail to one corner of the triangular tube with about $\frac{1}{2}$ inch of the nail sticking out beyond the edge. You should be able to rotate the circles freely on the nail.
9. Hold the tube close to but not touching your eye, and look through the open end while you rotate the circles.

What's Going On Here

Putting duct tape on the pieces of Plexiglas makes them behave like mirrors. The beautiful patterns that are produced by the spinning circles are due to the multiple reflections coming off the three Plexiglas mirrors.

RADICAL REFLECTIONS

What You Want to Know

What do you see when you look through a triangular tube of mirrors?

What You Think Will Happen

When you look through a triangular tube of mirrors at objects in a room, you will see

>   a. many reflections of your eye.
>   b. many reflections of objects in the room.
>   c. three reflections of your eye.
>   d. three reflections of objects in the room.

When you spin acetate circles with designs at the end of a triangular tube of mirrors and look through the tube, you will see

>   a. three reflections of the designs.
>   b. many reflections of the designs.
>   c. three reflections of your eye.
>   d. many reflections of your eye.

What Happened

Describe what you saw when you looked through the triangular tube of mirrors at objects in the room.

_____

Describe what you saw when you looked through the triangular tube of mirrors with the acetate design spinning at the end of the tube.

_____

What It Means

What do your observations tell you about what happens when light is reflected off three mirrors in a triangular tube?

_____

You placed duct tape on pieces of Plexiglas to make mirrors. What do your observations tell you about how reflections are different using plain Plexiglas and Plexiglas with duct tape on it?

_____

REALLY RADICAL REFLECTIONS

 SCIENCE: Multiple images are produced when light from an object reflects off several mirrors.

STUFF: Two pieces of duct tape (each about 2 inches wide and 6 inches long), three pieces of Plexiglas (1 inch x 6 inches), scissors, plastic wrap, small beads (or sequins), wax paper, rubber band

What to Do

1. Lay one piece of duct tape on a table, sticky side up.
2. Place one piece of Plexiglas flat in the middle of the tape, going lengthwise.
3. Place the second and third pieces of Plexiglas with their long edges touching the long edges of the first piece. Bend the second and third pieces inward so that their upper long edges touch each other. Pull the excess tape along the edges of the first piece up so that the tape holds the second and third pieces of Plexiglas in place. The short sides of the three pieces of Plexiglas should form a triangle.
4. Lay the long side of a second piece of duct tape along the long side of the second piece of Plexiglas, just overlapping the first piece of duct tape. Fold the duct tape around the uncovered pieces of Plexiglas so that the triangular tube is completely covered with duct tape.
5. Cut a piece of plastic wrap slightly larger than the end of the triangular tube. Place the plastic wrap over one end of the tube, and push it slightly into the tube to form a bowl. Add a few beads (not more than seven) to the bowl.
6. Cut a piece of wax paper slightly larger than the end of the triangular tube. Place the wax paper over the end of the tube with the plastic wrap. Pull it taut, and secure it with a rubber band around the tube. This is your completed kaleidoscope.
7. Hold the open end of the kaleidoscope close to but not touching your eye, and look through it at the beads or sequins. Shake the kaleidoscope gently to change the patterns.

What's Going On Here

Putting duct tape on the pieces of Plexiglas makes them behave like mirrors. The beautiful patterns that are produced by the beads are due to the multiple reflections coming off the three Plexiglas mirrors.

Try It! Try placing different objects in the kaleidoscope. Since the wax paper is easily removed, you can experiment with all sorts of objects.

REALLY RADICAL REFLECTIONS

What You Want to Know

What do you see when you look through a triangular tube of mirrors that has beads trapped under a layer of wax paper at one end?

What You Think Will Happen

When you look through a triangular tube of mirrors with beads trapped under a layer of wax paper at the other end, you will see

a. many reflections of the beads.
b. many reflections of your eye.
c. three reflections of your eye.
d. three reflections of beads.

What Happened

Describe what you saw when you looked through the triangular tube of mirrors with beads at the other end.

_____

What It Means

What can you now say about what happens when light is reflected off three mirrors in a triangular tube?

_____

Opaque objects do not let light pass through them. A sheet of cardboard is opaque. List all the materials used in this activity that are opaque.

_____

Transparent objects let light pass through them, so you can see objects clearly behind them. Glass is transparent. List all the materials used in this activity that are transparent.

_____

Translucent objects let some light pass through them, but you can't see objects clearly that are behind them. A piece of tissue paper is translucent. List all the materials used in this activity that are translucent.

_____

CONCAVE, CONVEX

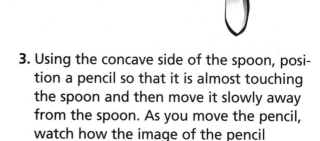

SCIENCE: Curved mirrors produce images that can be smaller or larger than the object. The images can also be upside down or right side up.

STUFF: Shiny large metal spoon, pencil

What to Do

1. Look into the concave side of the spoon—that is, the side where food rests. Notice how large your image is and whether it is upside down or right side up.
2. Look into the convex side of the spoon—that is, the back side of the spoon. Notice how large your image is and whether it is upside down or right side up.

3. Using the concave side of the spoon, position a pencil so that it is almost touching the spoon and then move it slowly away from the spoon. As you move the pencil, watch how the image of the pencil changes in size and orientation.
4. Repeat Step 3 for the convex side of the spoon.

What's Going On Here

The images from the concave and convex mirror sides of the spoon vary in size and orientation depending on the distance between the object and the mirror. A convex mirror is straightforward. The image from a convex mirror is always smaller than the object and is always right side up. The image from a concave mirror is more complex. When an object is very close to the mirror, the image is right side up and larger than the object. As the object is moved away from the mirror, it suddenly flips upside down and at first is still larger than the object. The image gets smaller as the object is moved farther from the mirror. When it is far enough from the mirror, the image will be upside down and smaller than the object.

Try It!

Try observing your image in a flat mirror. Notice how the image changes as you move closer to the mirror and then farther away.

Try other sizes of spoons. Try to find a very large serving spoon.

Find other examples of concave and convex mirrors.

Use a shiny round Christmas ornament, and observe your image in it.

Place several shiny Christmas ornaments together on a flat surface, and look at your image in the array.

CONCAVE, CONVEX

What You Want to Know

What does your reflection look like when you look into each curved side of a shiny spoon?

What You Think Will Happen

When you look into the side of a spoon that your food rests in (the concave side), your reflection will appear

    a. upside down and larger.
    b. upside down and smaller.
    c. right side up and larger.
    d. right side up and smaller.

When you look into the side of a spoon that your food does not rest in (convex side), your reflection will appear

    a. upside down and larger.
    b. upside down and smaller.
    c. right side up and larger.
    d. right side up and smaller.

What Happened

What was your reflection like when you looked into the concave side of the spoon?

_____

What was your reflection like when you looked into the convex side?

_____

What did the reflection of the pencil look like when the pencil was almost touching the concave side of the spoon?

_____

What did the reflection of the pencil look like when the pencil was almost touching the convex side?

_____

What It Means

What can you now say about a reflection in a curved mirror?

_____

_____

© Dale Seymour Publications®

LIQUID LENS

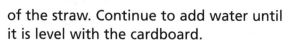

SCIENCE: A lens can be made by making a large water drop.

STUFF: Cardboard (about 6 inches x 6 inches), quarter, pencil, scissors, plastic wrap, tape, straw, cup of water, newspaper, ruler

What to Do

1. Place a quarter in the middle of a piece of cardboard, and draw a circle around it. Cut the circle out.
2. Put a piece of plastic wrap over the hole. Push the plastic into the hole just enough to make a noticeable indentation (about $\frac{1}{8}$ of an inch). Tape the edges of the plastic onto the cardboard.
3. Place a straw into a cup of water. Put your index finger over the top of the straw, and remove the straw from the water. Hold the straw over the plastic-covered hole in the cardboard (taped side up), and take your finger off the top of the straw. Continue to add water until it is level with the cardboard.
4. Hold the cardboard horizontally about an inch above a newspaper article, and look through the water at the newsprint below. Vary the distance of the cardboard from the newspaper until you see a magnified image of the newsprint.
5. Move the cardboard farther from the newsprint until the image of the newsprint appears upside down. At that point, measure the distance from the cardboard to the newspaper.

What's Going On Here

When water is placed on the plastic, it takes the shape of a plano-convex lens, flat on the top and curved on the bottom. This kind of lens can be used as a magnifier. It bends the light rays coming from the newsprint so that the newsprint appears to be closer than it actually is.

Try It!

Try drawing a circle on a microscope slide with a crayon. Put a drop of water in the middle of the circle, and hold the lens over something you would like to magnify.

Try magnifying other objects.

Try using other liquids besides water.

LIQUID LENS

What You Want to Know
How can you make a magnifying glass (lens) out of water?

What You Think Will Happen
When you put water onto plastic placed over a hole in a piece of cardboard, the water takes a rounded shape on the bottom and a fairly flat shape on the top. If you hold the large water drop about an inch above a newspaper, the newsprint will appear

       a. larger and right side up.
       b. larger and upside down.
       c. smaller and right side up.
       d. smaller and upside down.

When you move the water drop farther from the newspaper, the newsprint will appear

       a. larger and right side up.
       b. larger and upside down.
       c. smaller and right side up.
       d. smaller and upside down.

What Happened
What happened when you held the water drop about an inch from the newsprint?

_____

What happened when you moved the water drop farther away?

_____

How far did you have to move the water drop before the image flipped?

_____

What It Means
What do your observations tell you about how a magnifying glass (lens) can be made using water?

_____

What shape is the bottom of the water drop? What shape is the top?

_____

MEASURING MAGNIFICATION

SCIENCE: The magnification of a lens can be determined by measuring how much bigger it makes an object appear.

STUFF: Magnifying glass, piece of lined notebook paper, several books

Lines outside lens = 8
Lines inside lens = 5
Magnification = $\frac{8}{5}$

What to Do

1. Place a magnifying glass (a kind of lens) on a piece of lined notebook paper. Count the number of lines that the glass covers from the top to the bottom of the lens. That number is the *lines outside lens* and does not change throughout this activity.

2. Hold the lens a few inches above the piece of notebook paper. The lines seen through the lens should appear farther apart.

3. Prop books under the lens to hold it in place. Measure the height of the magnifying glass from the paper.

4. Count the number of lines of notebook paper that lie within the outline of the lens. That number is the *lines inside lens,* and it will vary, depending on how far from the paper the lens is.

5. Divide the number of lines outside the lens (Step 1) by the number of lines inside the lens (Step 4). This is the magnification of the lens.

6. Change the distance of the lens from the paper, and repeat Steps 3–5.

What's Going On Here

Converging lenses are thicker in the middle than at the edges. The focal length of a converging lens can be determined by holding the lens under a ceiling light and moving the lens up and down from a piece of paper placed on the floor. When the ceiling light is in focus on the paper, the distance from the paper to the lens is the focal length of the lens. This will work provided that the ceiling light is at least 10 times the distance from the lens as the focal length of the lens. Once you know the focal length of the lens, you need only be closer to the object you want to magnify than the focal length for the object to look larger. The amount of magnification will vary with how close to the paper you are; the closer to the paper you get, the smaller the magnification will be.

Try drawing extra lines on your notebook paper so that there are more lines per inch and you can get a more accurate number for the magnification.

Try using a different magnifying glass.

Try using the "liquid lens" described in the previous activity.

MEASURING MAGNIFICATION

What You Want to Know

How can you measure how much bigger a magnifying glass (lens) makes an object appear? How does moving a lens farther from an object affect the magnification of the lens (how much bigger it makes the object look)?

What You Think Will Happen

When you hold a lens a few inches from a piece of lined paper,

a. fewer lines will be seen inside the lens than outside it.
b. more lines will be seen inside the lens than outside it.
c. the same number of lines will be seen inside and outside the lens.

As you move a lens farther from a piece of lined paper, the lines on the paper will

a. not change.
b. get closer together.
c. get farther apart.

What Happened

Record your observations in the table.

Position of lens	Lines outside lens	Lines inside lens	Magnification = lines outside lens / lines inside lens
Flat on paper			
____ inches above paper			
____ inches above paper			

What It Means

What can you now say about how magnifying a piece of notebook paper changes the number of lines you see inside a lens?

_____

What can you now say about how moving a lens farther from a piece of notebook paper affects the magnification?

_____

INVERTED IMAGES

SCIENCE: The location of an image formed by a converging lens depends on the focal length of the lens and the location of the object.

STUFF: Black construction paper, scissors, pencil, ruler, tape, wax paper, unlined index card (3 inches x 5 inches), wooden block (2-inch cube or larger), flashlight, magnifying glass, clay

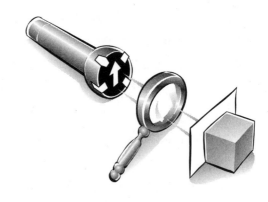

What to Do

1. Out of a piece of black construction paper, cut a circle the size of the front part of a flashlight. Draw an arrow about $\frac{1}{8}$ inch thick on the circle, and cut it out. Tape a piece of wax paper onto the circle just large enough to completely cover the arrow. Tape the black circle to the front of the flashlight. Record the height of the arrow.

2. Determine the focal length of a magnifying glass by placing an index card on the floor. Move the magnifying glass up and down until you find an image of the overhead lights on the index card. Record the distance from the index card to the magnifying glass. This is the focal length of the lens. Note that this procedure will work well only if the distance from the magnifying glass to the light is at least 10 times the focal length of the lens.

3. Tape the index card to a face of a wooden block.

4. In a darkened room, lay the flashlight horizontally on a table so that the arrow is pointing up. Turn the flashlight on. Place the lens on its side on the table at about twice its focal length from the arrow on the flashlight. If the lens cannot balance by itself, use clay to hold it in place.

5. Position the wooden block with the index card facing the lens and flashlight. Move the block away from and toward the lens until you find a sharp image of the arrow. Record the distance from the arrow to the lens; this is called the object distance. Record the distance from the lens to the index card; this is the image distance. Also record the image height.

6. Repeat Step 5 for four other locations of the object. (The object distance should not be less than the focal length of the lens.)

What's Going On Here

A converging lens is thicker in the middle than at the edges. The focal length of a converging lens is the distance from the lens to a point where a distant object is focused. If an object is located at a distance beyond the focal length of the lens, an upside-down image can be located at some point on the other side of the lens from the object. As the object distance is increased, the image distance is decreased and the image size is decreased.

INVERTED IMAGES

What You Want to Know
Where is an image from a magnifying glass (lens) located? What does the image look like?

What You Think Will Happen
As you move an object farther from a magnifying glass, the image will
   a. move farther from the lens.
   b. move closer to the lens.
   c. not move.

As you move an object farther from a magnifying glass, the image will
   a. become smaller.
   b. become larger.
   c. not change in size.

What Happened
The height of the arrow on the flashlight was _____.
The focal length of the lens was _____.

Record your observations in the table.

Try	Object distance	Image distance	Image height
1			
2			
3			
4			
5			

What It Means
What do the numbers in your table tell you about how the object distance and the image distance are related?

_____

What do your observations tell you about the appearance of the image?

_____

© Dale Seymour Publications®

SIMPLY SYMMETRICAL

BOB HID MARY'S TOY

SCIENCE: A cylindrical lens held close above words inverts them. If the words have horizontal symmetry, they will appear unaffected.

STUFF: Test tube with cork (or clear plastic coin tube, penny size), water, piece of white photocopy paper, blue and red markers

What to Do

1. Fill a test tube fully with water. Place a cork tightly in the top. You have just made a cylindrical lens.
2. On a piece of paper, write the words BOB HID with a blue marker. Just next to that, write MARY'S TOY with a red marker. Try to make all the letters the same size, and keep them in line. Also be sure the sentence is shorter than the length of the tube.
3. Hold the tube horizontally over the words and a few inches above them. Look at the words through the tube. The tube appears to invert the red letters but not the blue letters.
4. Try other combinations of letters to solve the mystery.

What's Going On Here

Letters that have horizontal symmetry appear the same when they are inverted over a horizontal line going through the middle of the letter. O, B, and C are examples of letters with horizontal symmetry. Letters that have vertical symmetry appear the same when they are flipped over a vertical line going through the middle of the letter. A, H, and O are examples of letters having vertical symmetry. The cylindrical lens inverts all the letters (makes them upside down). Letters with horizontal symmetry will appear the same after being inverted as they appeared before. BOB HID is actually inverted, but since all the letters have horizontal symmetry, they appear the same after being inverted. MARY'S TOY is inverted too, but since most of these letters do not have horizontal symmetry, they appear different after being inverted.

Try It!

Try to use the cylindrical lens to determine which letters in the alphabet have horizontal symmetry.

Try to use the cylindrical lens to determine which letters in the alphabet have vertical symmetry. You will have to change the direction that you hold the cylindrical lens.

Try drawing other shapes with vertical or horizontal symmetry and looking at them through the cylindrical lens.

SIMPLY SYMMETRICAL

What You Want to Know

When a clear tube filled with water is held just above some words, what do the words look like through the tube?

What You Think Will Happen

When you look through a clear tube filled with water held just above the words BOB HID MARY'S TOY,

     a. the words will be bigger.

     b. the words will be smaller.

     c. some words will be upside down.

     d. all the words will be upside down.

     e. _____.

What Happened

Write down exactly what you saw when you held the water-filled tube over the words BOB HID MARY'S TOY.

_____

Write down the other letter combinations that you tried and exactly what you saw.

_____

_____

What It Means

Which of the printed capital letters of the alphabet appear upside down through a water-filled tube?

_____

What do the letters that do not appear upside down when viewed through a water-filled tube have in common?

_____

SUPER SPECTRUM SPECTACLES

SCIENCE: Light reflected from objects often contains many different wavelengths, or colors, of light.

STUFF: Tagboard (or thin cardboard), scissors, blank sheets of acetate (or transparencies or report covers) in various colors, tape

What to Do

1. From a piece of tagboard, cut a pair of eyeglass frames.
2. Cut two pieces of the same color acetate to fit the eyeglass lens areas. Tape them in place.
3. Look at objects around the room. Note the color that the object actually is and the color that it appears when you have the colored glasses on.
4. Change the color of the eyeglass lenses, and repeat Step 3.
5. Repeat Step 3 for a third color of lenses.

What's Going On Here

We see most objects by reflected light. Usually the light shining on the objects is white light, which contains all the colors of the spectrum, or rainbow. When white light strikes an object, the different colors of the white light are absorbed, reflected, or transmitted by the object. The light we see when we look at the object directly is reflected light. The various colors of reflected light are combined to give us the sensation of a certain single color of light. An object that appears red may be reflecting all the colors to a certain degree, but the colors combine in such a way as to make the object appear red. The colored glasses you create in this activity make some objects look different than they do in normal light. That is because the colored acetate allows only certain colors of light to pass through it. When you look at an object with the eyeglasses, your eyes combine only the colors that are reflected from the object *and* allowed to pass through the colored lenses.

Try It!

Try looking at magazine pictures.

Try coloring a picture with markers or crayons while wearing the glasses. Then take the glasses off, and look at the picture.

Try putting a different color of transparency on each lens and then looking at objects around the room.

SUPER SPECTRUM SPECTACLES

What You Want to Know
How do the colors of objects change when you look at them through colored eyeglasses?

What You Think Will Happen
When you look at an object through colored eyeglasses,

      a. its color will always look the same.
      b. its color will never look the same.
      c. its color will sometimes look the same.

What Happened
In the table below, list the colors of eyeglass lenses that you used. Also list the name and color of the objects that you looked at. Then write the color of each object as it appeared through each pair of glasses.

Object and color	Eyeglass lenses		
	Color _____	Color _____	Color _____

What It Means
What do your observations tell you about how the colors of objects change when they are looked at through colored eyeglasses?

_____

_____

RAVISHING RAINBOWS

SCIENCE: A rainbow is the dispersal of white light into its component colors. Different colors of light pass through different colors of acetate.

STUFF: Overhead projector, piece of aluminum foil, piece of white photocopy paper, tape, 5- or 10-gallon fish tank (or clear baking dish), water, blank sheets of acetate (or overhead transparencies or report covers) in three colors

What to Do

1. Put an overhead projector on a table in the center of the front of the room.
2. Tape a piece of aluminum foil over the upper mirror unit of the overhead projector. The activity will work with the mirror unit uncovered, but it is not needed and it can be distracting.
3. Place a small fish tank on the flat lighted top of the projector. Fill the tank about halfway with water. Darken the room.
4. An arced spectrum, or rainbow, should appear on two walls of the room. (In some rooms, you may have to look elsewhere for the rainbow.) Hold a sheet of white paper about a foot from the side of the tank to find where the rainbow is heading, and then follow it!
5. Place the colored sheets of acetate, one at a time, against the fish tank at the point where the rainbow is leaving the tank and heading for the wall. Notice which colors of the rainbow are present for each transparency.

What's Going On Here

The light from the projector lens (which is actually the flat surface that the fish tank is placed on) enters the sides of the container of water and is dispersed into colors by refraction. *Refraction* is the bending of light as it goes from one material to another—for example, from air to glass or glass to water. Light is refracted, or bent, differently for different wavelengths, or colors. White light has all the colors in the rainbow in it, so when white light from the projector is bent through the water and glass of the container, a rainbow is produced as the different colors are bent differing amounts. Red is bent less than yellow, which is bent less than green, which is bent less than blue, and so on. When you placed the colored transparency in front of the rainbow, parts of the spectrum are absorbed by the transparency and parts are transmitted. The parts that are transmitted are seen in what's left of the full rainbow on the wall. This is a good way of analyzing which colors are actually transmitted by a given color of transparency. A transparency that looks yellow to our eyes, for example, may actually transmit the green and orange parts of the spectrum, too. Our eyes combine the colors that are transmitted by the transparency in such a way that the transparency looks yellow.

RAVISHING RAINBOWS

What You Want to Know
What colors of light pass through different colors of acetate sheets?

What You Think Will Happen
A tank of water placed on an overhead projector is used to make a rainbow. Different colors of acetate sheets are placed against the tank of water so that only the colors that can pass through the acetate show up in the rainbow. In the table below, list the colors of acetate sheets that you will use; then predict what colors you will be able to see in the rainbow for each sheet.

Color of acetate	Colors in rainbow

What Happened
In the table below, list the colors of acetate sheets you used. Then write the colors you saw in the rainbow for each sheet.

Color of acetate	Colors in rainbow

What It Means
What do your observations tell you about what colors pass through different colors of acetate sheets?

_____

COLOR COMBINATIONS

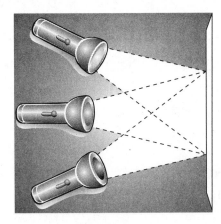

SCIENCE: Colored lights can be mixed together to make other colors.

STUFF: Scissors; blank sheets of acetate (or overhead transparencies or report covers) in red, blue, and green; three flashlights; black tape; white screen (or large sheet of white paper); ruler

What to Do

1. Cut a circle from each color of transparency to fit over the front of a flashlight. Tape each on a flashlight. Use black tape, and cover any exposed areas so that light goes through the transparencies only.
2. Set the flashlights on their side on a table. Then darken the room. Aim the flashlights at a single point on a screen.
3. Make shadows on the screen.
4. Hold a ruler just in front of the screen, and record the colors you can see in the shadow of the ruler.
5. Turn off one flashlight at a time, and again make shadows. Record the colors of the shadows you see when you hold the ruler just in front of the screen.
6. Turn off two flashlights at a time, and again make shadows. Record the colors of the shadows you see when you hold the ruler just in front of the screen.

What's Going On Here

Red, blue, and green are the primary colors for additive color mixing, which works as follows: Red, blue, and green lights combine to give white light. When you make a shadow in front of these lights, the color of the shadow depends on which of the three lights you block out to make the shadow. Areas on the screen where you block out the red light will appear cyan because of the mixing of the green and blue lights. Yellow will appear where you block out the blue light because the red and green lights combine to give yellow. The shadow will be magenta where you block out the green light; red and blue combine to give the magenta color. There are some places where you can block out two colors at the same time. The shadow will appear red where you block out the green and blue lights, since red is all that is left. The shadow will appear blue where you block red and green, and the shadow will appear green where you block the blue and red lights. Of course where you block all the lights, the shadow will be black. With the combinations of the three lights, it is possible to create shadows of six different colors (cyan, magenta, yellow, green, blue, and red) plus black and white.

Try It! Try using other colors of transparencies.

COLOR COMBINATIONS

What You Want to Know

What do shadows look like when the lights used to make them are red, blue, and green in various combinations?

What You Think Will Happen

When you make a shadow on a screen where red, blue, and green lights are aimed at one spot, you will see

a. just black.

b. just red, blue, and green.

c. red, blue, green, and other colors.

What Happened

What colors of shadows did you see when you used the combinations of lights listed in the table?

Color of lights	Color of shadows
Red, blue, green	
Red, blue	
Red, green	
Green, blue	
Red	
Blue	
Green	

What It Means

You are making shadows using red, blue, and green lights. When you block the red light, the blue and green lights mix together to make a color called cyan (a blue green). What do your observations tell you about the results when the other colors of light mix together?

_____

_____

© Dale Seymour Publications®

SHIMMER AND SHINE

SCIENCE: White light reflecting from a thin film produces a rainbow.

STUFF: Newspaper, piece of black construction paper (9 inches x 12 inches), scissors, tape, string, water, disposable metal pie pan, clear nail polish

What to Do

1. Cover your work area with newspaper.
2. Cut a piece of construction paper into fourths, and tape a short piece of string to the edge of one of these pieces of construction paper.
3. Put about $\frac{1}{2}$ inch of water in the pie pan.
4. Slide the paper into the water, being careful to let the string hang outside the edge of the pie pan.
5. Let one drop of nail polish fall from the nail polish brush into the water. It should spread out over the surface of the water. Look at the reflection of light in the thin layer of nail polish on the water.
6. Using the string, slowly pull the paper *up and out* of the water. A thin layer of nail polish should stick to the paper.
7. Hold the paper at an angle, and move it around under a light to see the rainbow of colors reflecting from its surface.
8. Allow the paper to dry, and look at the reflection again. Dispose of the pie pan.

What's Going On Here

A thin layer of nail polish is deposited on the surface of the black paper. When light reflects off the nail polish, it is actually reflecting *twice,* once off the top of the layer of nail polish and once off the bottom of the layer of polish. The light reflecting from the top and from the bottom of the layer of polish interfere with each other to produce the rainbow of colors that appear in streaks and fringes on the paper.

Try It!

Try using a thin layer of oil on water.

Try using some other color of construction paper, including white.

Try brushing nail polish on a piece of construction paper.

SHIMMER AND SHINE

What You Want to Know
What does the reflection from a thin layer of nail polish on construction paper look like?

What You Think Will Happen
When you look at the reflection from a thin layer of nail polish on a piece of black construction paper, you will see

       a. a shiny surface, but no colors.
       b. a shiny surface with many colors.
       c. a soggy piece of black construction paper.
       d. _____.

What Happened
Describe the reflection from the thin layer of nail polish on top of the water in the pie pan.

_____

Describe what you saw when you looked at the thin layer of nail polish on the wet black construction paper.

_____

Describe how the reflection changed when you looked at the paper after it had dried.

_____

What It Means
What can you now say about what the reflection from a thin layer of nail polish looks like?

_____

You may have seen a thin layer of gas or oil on a puddle on a road. How is that similar to what you saw in this activity?

_____

CHAPTER 7 | AIR PRESSURE

Practical Pressure Particulars

- The atmosphere, or blanket of air that surrounds the earth, pushes down on the earth because of the weight of the air; this effect is called air pressure, or atmospheric pressure.
- Air pressure varies with your location on the earth: it is about 15 pounds per square inch at sea level. As you get higher above sea level, the air pressure decreases because there is less air above you pushing down.
- Faster-moving air exerts less pressure against an object than slower-moving air. This concept was discovered by the Swiss mathematician David Bernoulli (1700–1782) and is part of Bernoulli's principle.
- Airplanes are able to fly because of Bernoulli's principle.
- Simply stated, there are four forces on a flying airplane:
  1. Thrust, the forward force on the airplane provided by the engines
  2. Drag, the force that opposes the forward motion of the plane; this is mostly a result of the force of air resistance
  3. Gravity, the weight of the airplane
  4. Lift, the force that opposes gravity; it is caused by the difference in air pressure between the top and the bottom surfaces of the airplane's wings.

- An airplane wing is designed so that air moves faster over the top of the wing than it does over the bottom of the wing. By Bernoulli's principle, the air pressure on the top of the wing is therefore less than it is on the bottom of the wing. Since the pressure on the bottom of the wing is greater than the pressure on the top of the wing, the wing is pushed up; this is called lift.
- The faster an airplane moves, the greater is the difference in air pressure between the top and the bottom of the wing, and the greater is the lift.
- If the lift is greater than the weight of the airplane, then the airplane will leave the ground and fly. If the weight of the airplane is greater than the lift, then gravity wins the tug-of-war with lift and the airplane will not fly.
- Water has pressure too. The deeper you go in a body of water, the more water is pushing down on you from above, and the greater the water pressure.

WILLFUL WATER

SCIENCE: Air pressure can keep water inside a bottle that has a hole below the water level.

STUFF: Scissors with pointed tip, plastic bottle with a screw-on cap, water, bucket

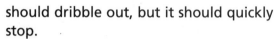

What to Do

1. Twist the pointed end of a pair of scissors to make a small hole on the side and near the bottom of a plastic bottle.
2. While covering the hole with your finger, fill the bottle with water and screw on the cap.
3. Hold the bottle over a bucket, and without squeezing the bottle, take your finger away from the hole. A little water should dribble out, but it should quickly stop.
4. While holding the bottle over a bucket, unscrew the cap. Water should pour from the bottle.
5. Continue to screw the cap on and off, using the bottle as a faucet. Notice the path the water takes as it streams from the bottle.

What's Going On Here

When the cap is screwed on the bottle, air pressure from outside the bottle presses on the water and holds it inside the bottle. A little dribbles out at first because of air pressure inside the bottle. As the water dribbles out, the air pressure inside the bottle becomes less because the air inside is occupying a larger space than before the water dribbles out. When the air pressure inside the bottle has become less than that outside the bottle, no more water will drain out. When you take the cap off the bottle, water streams out the hole because the force of air pressure on the top of the water surface is the same as the air pressure on the side of the bottle and gravity pulls the water out of the bottle.

Try It!

Try squeezing the bottle when the cap is screwed on.

Predict how far the water will squirt out from the hole for different heights of water in the bottle.

Try poking a hole in the bottom of the bottle.

WILLFUL WATER

What You Want to Know
A water-filled bottle has a small hole in the side near the bottom. What happens as you loosen and tighten the cap on the bottle?

What You Think Will Happen
When you take the cap off a water-filled bottle with a hole near the bottom,

        a. the water will stream out the hole.
        b. the water will dribble out the hole and quickly stop.
        c. no water will come out the hole.

When you tighten the cap of the bottle,

        a. the water will stream out the hole.
        b. the water will dribble out the hole and quickly stop.
        c. no water will come out the hole.

What Happened
What happened when you took the cap off the bottle?

_____

What happened when you tightened the cap of the bottle?

_____

Describe the path of the water as it streamed from the bottle at first.

_____

Describe the path of the water as the bottle became almost empty.

_____

What It Means
What can you now say about what happens when you loosen and tighten the cap of a water-filled bottle with a hole near the bottom?

_____

What is pushing down on the water when it streams from the bottle?
What is keeping the water in the bottle when it does not stream out?

_____

TRICKY TUG-OF-WAR

SCIENCE: Air pressure is a force that can compete with the force of gravity.

STUFF: Clear plastic cup (8 or 10 ounces), scissors with pointed tip, water, piece of thin cardboard large enough to cover the top of the cup, bucket

What to Do

1. Poke a small hole near the bottom of a plastic cup, using the tip of a pair of scissors. The hole should be about $\frac{1}{8}$ inch in diameter.
2. While holding a finger over the hole, fill the cup with water.
3. Still covering the hole, place a piece of cardboard on top of the cup of water. Move the cup around a little so that the cardboard gets wet and forms a good seal with the rim of the cup.
4. Hold the cardboard tightly against the top of the cup, and carefully invert the cup. Do this over a bucket, and make sure you still have a finger over the hole in the cup.
5. While holding the cup over the bucket, take your hand away from the cardboard so that you are holding only the cup (with one finger still over the hole). The water should remain in the cup.
6. Pull your finger away from the hole in the cup. The water should rush out, to the amazement of onlookers.

What's Going On Here

Gravity is pulling the water inside the inverted cup down with a force equal to the weight of the water. When you hold your finger over the hole, air pressure is pushing up on the thin cardboard with a force that is greater than the weight of the water and the water stays in the cup. In other words, air pressure wins the tug-of-war, and the water is held in the cup. When you pull your finger away from the hole in the cup, the air pressure inside the cup equals the air pressure outside the cup. Gravity acting on the water wins the tug-of-war against air pressure, and the water rushes out!

Try It!

Try different amounts of water in a cup without a hole in it.

Try putting the hole near the top of the cup.

Using a cup without a hole in it, try squeezing the cup when you hold it upside down.

Try using a glass without a hole in it.

TRICKY TUG-OF-WAR

What You Want to Know

What happens when you place a piece of cardboard on a cup of water and then turn the cup and cardboard upside down? What happens if there is a small hole near the bottom of the cup when you turn it upside down?

What You Think Will Happen

When you place a piece of cardboard on top of a cup of water and then turn the cardboard and cup upside down together,

> a. the water will stay in the cup.
> b. the water will spill out of the cup.

When you place a piece of cardboard on top of a cup of water with a covered hole near the bottom, turn the cardboard and cup upside down, and open the hole,

> a. the water will stay in the cup.
> b. the water will spill out of the cup.

What Happened

What happened when you turned the cardboard and cup upside down while holding your finger over the hole near the bottom of the cup?

_____

What happened when you turned the cardboard and cup upside down and pulled your finger away from the hole near the bottom of the cup?

_____

What It Means

What can you now say about what happens when you turn a water-filled cup with a piece of cardboard on it upside down together?

_____

What is pushing up on the cardboard to keep the water in the cup?

_____

What is pushing down on the water when it spills out of the cup?

_____

BREATHING BALLOON

 SCIENCE: Contracting and relaxing of our diaphragm causes our lungs to take in and push out air.

STUFF: Clear plastic cup (8 or 10 ounces), electric drill and $\frac{3}{8}$-inch bit, rubber bands, one small balloon (12 inches or less), straw, clay, one large balloon (15 inches)

What to Do

1. Turn a clear plastic cup upside down, and hold it firmly as you drill a hole through the middle of the bottom of the cup. (Be careful as you use the drill.)
2. Using a rubber band, attach a small balloon to one end of the straw. Blow through the other end of the straw. You should be able to partially inflate the balloon.
3. Put the straw through the hole in the plastic cup, and seal the hole with clay. The balloon should be totally inside the cup.
4. Cut off the top end of a large balloon a third of the way down from the blowing end. Stretch the rest of the balloon over the top of the cup, and secure it with a rubber band.
5. Gently push the stretched balloon inward toward the inside of the cup. The balloon inside should partially deflate.
6. Gently pinch the middle of the stretched balloon, and pull it outward from the cup. The balloon inside should partially inflate.

What's Going On Here

When the stretched balloon is pushed inward, air pressure inside the cup increases and air is pushed out of the inner balloon through the straw. When the stretched balloon is pulled outward, air pressure inside the cup decreases and air from outside the cup is drawn into the inner balloon. This activity shows how the diaphragm works. The stretched balloon is like the diaphragm. The small balloon is analogous to a lung, and the straw is like a windpipe. When the diaphragm contracts, the chest cavity is expanded and the resulting decreased air pressure in the lungs causes air to be drawn in through the windpipe via the nose or mouth. When the diaphragm relaxes, the chest cavity is contracted and the resulting increased air pressure causes air to be exhaled.

Try It!

Make the hole in the cup a little bigger, and add a second straw and balloon to simulate two lungs.

Poke a small hole in the stretched balloon, and try the activity.

Poke a small hole in the small balloon, and try the activity.

BREATHING BALLOON

What You Want to Know
How do we breathe?

What You Think Will Happen
A small balloon attached to a straw is placed inside a
cup that has a larger balloon stretched over the rim
like a drum. When you push inward on the larger balloon,

      a. the smaller balloon will deflate (get smaller).
      b. the smaller balloon will inflate (get larger).
      c. the smaller balloon will stay the same.

When you pull outward on the larger balloon,

      a. the smaller balloon will deflate (get smaller).
      b. the smaller balloon will inflate (get larger).
      c. the smaller balloon will stay the same.

What Happened
What happened when you pushed inward on the larger balloon?

_____

What happened when you pulled outward on the larger balloon?

_____

What It Means
Name the parts of the breathing machine you made that are like a lung,
a windpipe, and a diaphragm.

_____

Pushing inward on the larger balloon caused the air inside the cup to
squeeze together, making the air pressure inside the cup greater. What
can you now say about the air pressure around your lungs when you
exhale (push air out of your lungs)?

_____

What can you now say about the air pressure around your lungs when
you inhale (pull air into your lungs)?

_____

CRUSHED CAN

SCIENCE: Difference in air pressure can crush a can.

STUFF: Empty soft drink cans, cold tap water, hot plate (or electric stove), bowl, tongs, insulated gloves (or pot holders)

What to Do

1. Twist the tab off the top of a soft drink can.
2. Place about $\frac{1}{4}$ inch of water in the can.
3. Place the can on a hot plate.
4. While you are waiting for the water in the can to boil, put 1 or 2 inches of cold water in a bowl.
5. After the water in the can has boiled for about a minute, remove it from the hot plate, using tongs and insulated gloves. Note: The can becomes very hot; be sure to handle it carefully. The water in the can is also very hot, so be careful not to splash it.
6. Immediately (without splashing any of the boiling water on yourself) turn the can upside down in the cold water as far as it will go.
7. If the can does not crush right away, try another soft drink can. Some cans are more stubborn than others!
8. Still using the insulated gloves, pull the can slowly out of the water, and notice how much water comes out of it.

What's Going On Here

When you turn the can upside down in the cold water, it immediately crushes. It does so because air pressure is greater on the outside of the can than on the inside of the can. When you boil the water in the can, water vapor is produced and it pushes most of the air out of the can. When the can is placed upside down in cold water, the water vapor condenses on the inside of the can in the form of water drops. The drops of water take up much less space than the water vapor, thus lowering the air pressure inside the can. If you take the can slowly out of the water, you will notice that a lot more water comes out of the can than you originally put in. This is because the air pressure outside the can was pressing on the water in the dish and pushing it upward into the area of lower pressure inside the can. Interestingly, the water can not be pushed up into the can fast enough to keep it from crushing as a result of the very quick condensation of the water vapor. Air pressure is actually acting in two ways in this activity, pushing water up into the can and crushing the can itself.

Try It!

Try using ice water instead of cold tap water in the bowl.

Try using hot tap water in the bowl.

Carefully measure how much water you put into the can. Then measure the amount of water that comes out of the can when you remove it from the bowl of cold tap water.

CRUSHED CAN

What You Want to Know
Can the force of air pressure be used to crush a can?

What You Think Will Happen
When you place a soft drink can with a small amount
of boiling water in it upside down in a bowl of cold water,

       a. the boiling water bubbles out of the can.

       b. the can will crush slightly.

       c. the can will totally collapse.

       d. the can will crush and then return to its original shape.

When you pull the can slowly out of the water, you will notice

       a. more water coming out of the can than you put in it.

       b. less water coming out of the can than you put in it.

       c. about the same amount of water coming out as you put in.

What Happened
What happened when you placed the soft drink can with the boiling
water in it upside down in the bowl of cold water?

_____

Compared with the amount of water you put in the can, how much
came out when you pulled the can out of the water?

_____

What It Means
Air pressure is the force of the air pushing on something. In this activity,
what do your observations tell you about the air pressure inside and
outside the can before you place it in a bowl of water?

_____

What do your observations tell you about the air pressure inside and
outside the can after you place the can with a small amount of boiling
water in it in a bowl of cold water?

_____

What causes the difference in air pressure inside and outside the can?

_____

BASHED BOTTLE

SCIENCE: Hot air takes up more space than cold air. Difference in air pressure can crush a bottle.

STUFF: Two bowls; ice; hot, warm, and cold tap water; plastic 2-liter soft drink bottle with cap

What to Do

1. Fill a bowl about one-fourth full of ice. Add cold water to make it half full.
2. Fill another bowl one-half full of warm water.
3. Pour hot water into a soft drink bottle until it is half full. Swish the water around in the bottle for about a minute.
4. Pour the water out, and quickly screw the bottle's cap on tightly.
5. Watch the bottle for a few minutes. The bottle should start to contract.
6. Place the bottle in the bowl of ice water.
7. Take the bottle out of the ice water, and place it in the bowl of warm water.

What's Going On Here

When you swish the hot water around in the bottle, it causes the air in the bottle to heat up and expand. Some of the hot air leaves the bottle. After you put the cap on, the bottle quickly cools down to room temperature and the air inside the bottle contracts as the particles of air move closer together. Water vapor condenses on the inside of the bottle as water drops. Since there are now fewer particles of air in the bottle and those particles are now closer together, the pressure inside the bottle is less than the air pressure outside the bottle and the bottle collapses.

Try the activity in reverse. Place the *open* bottle in a freezer. Take it out after a few minutes, and immediately put a balloon (instead of the cap) over the top. You should use a balloon because the bottle may expand so much that it will break if you use the screw-on cap.

Try different temperatures of water, and compare the results.

BASHED BOTTLE

What You Want to Know
What happens when you put the cap on a plastic bottle that has had hot water swished around in it and then dumped?

What You Think Will Happen
When a sealed bottle that had hot water swished around in it is put into ice water,

> a. the bottle will crush slightly.
> b. the bottle will totally collapse.
> c. the bottle will get slightly larger.

When you take the bottle from ice water and place it in warm water,

> a. the bottle will crush slightly.
> b. the bottle will totally collapse.
> c. the bottle will return to its original shape.

What Happened
What happened when a sealed bottle that had hot water swished around in it was put into ice water?

_____

What happened when you took the bottle out of the ice water and placed it in warm water?

_____

What It Means
Air pressure is the force of air pushing on something. In this activity, air is pushing on the inside and the outside of the bottle. What do your observations tell you about the air pressure inside and outside the bottle when the cap is off?

_____

What do your observations tell you about the air pressure inside and outside the bottle when you put the cap on and place it in a bowl of ice water?

_____

ELASTIC EGG

SCIENCE: Air pressure is a force that can push an egg into a bottle.

STUFF: Paper towel, matches, glass bottle (with mouth just wide enough for egg to sit on it), peeled hard-boiled egg, water

What to Do

1. Fold a paper towel the long way several times to make a long, narrow strip. Light one end of the towel with a match, and drop the towel into the bottle.
2. Put an egg on the bottle's mouth, with the narrow end pointed into the bottle.
3. Wait a minute or two. The egg should be pushed into the bottle.
4. Now for the tricky part—getting the egg out again! Pour some water into the bottle, and swish it around. Hold the egg away from the mouth of the bottle with your finger, and pour the water out, along with the ashes and burnt towel. Dry the mouth of the bottle.
5. Hold the bottle with its mouth pointing toward the floor. Move the egg around so that it is resting in the mouth of the bottle with its narrow end pointing out of the bottle.
6. With your lips *together,* put the mouth of the inverted bottle to your lips. *Do not wrap your lips around the mouth of the bottle,* or you will end up with a dirty egg in your mouth in the next step!
7. Blow into the bottle as hard as you can until you run out of breath. Then, keeping the bottle pointing downward, pull the bottle away from your lips. The egg should pop out.

What's Going On Here

When you place the egg on the mouth of the bottle with the burning paper towel inside, two things happen to lower the air pressure inside the bottle. The air inside the bottle is heated, so some of it escapes. You can see that happening if you carefully watch the egg as it vibrates on top of the bottle before it is pushed in. The other thing that lowers the air pressure inside the bottle is the combustion of the towel, which consumes oxygen and produces carbon dioxide and water vapor. As the water vapor cools, it condenses into drops of water on the inside of the jar. The drops of water take up much less space than the water vapor, thus lowering the air pressure inside the bottle. Since the air pressure inside the bottle is lower than the air pressure outside the bottle, the egg is pushed into the bottle. When you blow into the bottle in the second part of the activity, you make the air pressure inside the bottle greater than the air pressure outside the bottle, and the egg is pushed out.

Try It!

Use a very small water balloon instead of the egg.

Try heating the bottom of the bottle with a hair dryer instead of using the burning paper towel.

ELASTIC EGG

What You Want to Know

What happens to a peeled hard-boiled egg placed
on the mouth of a bottle after a piece of burning
paper has been put inside the bottle?

What You Think Will Happen

When you place a peeled hard-boiled egg on the mouth of a bottle
after a piece of burning paper has been put inside the bottle,

     a. the egg will pop off the top of the bottle.

     b. the egg will be pushed totally inside the bottle.

     c. the egg will be pushed slightly into the mouth of the bottle.

     d. _____.

What Happened

What happened when you placed the peeled hard-boiled egg on the
mouth of the bottle after a piece of burning paper had been put inside
the bottle?

_____

What happened when you blew into the bottle in the second part of
the activity?

_____

What It Means

Air pressure is the force of air pushing on something. In this activity, air
is pushing on the inside and the outside of the bottle. What can you
now say about the air pressure inside and outside the bottle when the
egg is resting on the mouth of the bottle?

_____

What can you now say about the air pressure inside and outside the
bottle when the egg is being pushed into the bottle?

_____

How does the temperature of the air inside the bottle change the
amount of air there? How does the temperature inside the bottle affect
the air pressure inside the bottle?

_____

UNBELIEVABLE UPLIFT

SCIENCE: Fast-moving air has less pressure than slow-moving air. The force of air pressure can compete with gravity.

STUFF: Funnel, table tennis ball, large marble

What to Do
1. Hold a funnel with its wide end pointing up.
2. Place the table tennis ball in the funnel.
3. Blow gently into the narrow end of the funnel. Then blow very hard.
4. Hold the funnel with its wide end pointing down. Use a couple of fingers to gently hold the ball in the wide end of the funnel against the opening to the narrow end.
5. Blow into the narrow end of the funnel as hard as you can. After a few seconds pull your fingers away from the ball.
6. Repeat Steps 1–5 with a large marble.

What's Going On Here
Fast-moving air has less pressure than slow-moving air. No matter how you hold the funnel when you blow into the stem, the air is moving faster over the part of the ball close to the stem than it is on the part of the ball close to the wide end of the funnel. The ball is pushed toward the stem because of the greater air pressure (and slower-moving air) on the part of the ball close to the wide end of the funnel. When the wide end of the funnel is pointed upward and you blow into it, gravity and air pressure work together to keep the ball in the funnel; both forces are pushing downward on the ball. When you hold the funnel so that the wide part is pointing downward and blow onto it, the ball stays in the funnel, seemingly defying gravity. In fact, gravity is very much a part of the story here; it is still a force pulling downward on the ball, but it is smaller than the force of air pressure pushing upward on the ball. Air pressure competes with gravity and wins—at least this time! If you try the marble or blow more gently with the table tennis ball, gravity may win.

Try It!

Try varying how hard you blow into the funnel.

Try attaching the stem of the funnel to a hair dryer, using duct tape. Run the dryer on "no heat."

Try different balls in the funnel.

Place the table tennis ball in the palm of your hand. Place the wide end of the funnel over the ball. See if you can blow hard enough into the funnel so that the ball rises from your hand.

UNBELIEVABLE UPLIFT

What You Want to Know
What happens when you blow into the end of a funnel with a table tennis ball sitting in the wide end? Will the results be different if the funnel is held upside down?

What You Think Will Happen
When you blow into the stem of a funnel with the stem pointing down,

    a. the ball will come out of the funnel and float in the air as long as you are blowing.

    b. the ball will come out of the funnel and spin around as long as you are blowing.

    c. the ball will stay snugly inside the funnel.

When you blow into the stem of a funnel with the stem pointing up,

    a. the ball will fall out of the funnel.

    b. the ball will come out of the funnel and spin around just below it.

    c. the ball will stay snugly inside the funnel.

What Happened
What happened when you blew into the funnel with its stem down?

_____

What happened when you blew into the funnel with its stem up?

_____

What It Means
The force of air pressure depends on the speed of the air. Faster-moving air has less pressure. In the drawing below, show the place on the ball that has faster-moving air on it (and so less air pressure). Explain how the difference in air pressure on the top and the bottom of the ball causes the ball to behave as it does.

_____

_____

_____

GRAPPLING GRAVITY

SCIENCE: Fast-moving air has less pressure than slow-moving air. The force of air pressure can compete with gravity.

STUFF: Thumbtack, index card (3 inches x 5 inches), thread spool, pencil

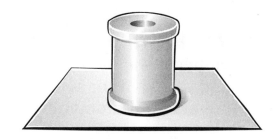

What to Do

1. Stick a thumbtack through the center of an index card.
2. If the spool you are using has paper on the ends, poke the paper out of the center of the spool, using the sharp end of a pencil. The top and the bottom of the spool should have an opening *only* in the center.
3. Place the index card against the spool, with the thumbtack positioned in the hole of the spool.

4. Hold the spool in one hand and the index card in the other so that the index card is horizontal and up against the bottom of the spool.
5. Blow into the top end of the spool as hard as you can. Once you have started blowing, pull your hand away from the card.
6. This activity takes some practice, but once you learn how to do it, you won't ever forget!

What's Going On Here

There are two forces on the index card, the force of gravity pulling it downward and the force of air pressure. The force of air pressure is the same on the top of the card as it is on the bottom of the card *until* you blow into the spool. Then something very interesting happens. Fast-moving air has less pressure than slow-moving air. Blowing through the spool causes fast-moving air to move over the top of the card creating lower air pressure on the top of the card.

Since there is less air pressure on the top of the card than on the bottom of the card, the card is pushed upward by air pressure. If this upward pressure is greater than the downward force of gravity, the card will stay up. If the upward air pressure is less than the force of gravity, the card will fall; this happens if you don't blow hard enough into the spool or if the card is too heavy.

Try It!

Try adding one index card at a time to see how many you can hold up against the spool by blowing through it.

Try a different size of card or spool.

Try the activity without the thumbtack in the index card.

Try varying how hard you blow into the spool.

GRAPPLING GRAVITY

What You Want to Know

A thumbtack is stuck through the center of a small card. A spool is placed on top of the card with its hole directly over the thumbtack. If you hold the card horizontally and blow into the spool, what happens to the card?

What You Think Will Happen

When you blow air into a spool and then let go of a card below,

a. the card will fall down.

b. the card will stay up against the bottom of the spool as long as you are blowing.

c. the card will stay up against the bottom of the spool even after you stop blowing.

What Happened

What happened when you blew air into the spool and then let go of the card below?

_____

What happened when you stopped blowing into the spool?

_____

What It Means

There are two forces at work on the card in this activity—gravity, which always pulls the card downward, and air pressure. When you blow into the spool, which force is greater?

_____

The force of air pressure depends on the speed of the air. Faster-moving air has less pressure. In the drawing below, make an arrow to show whether the top or the bottom of the card has faster-moving air on it (and so less air pressure). Explain how the difference in air pressure on the top and the bottom of the card causes the card to behave the way it does.

_____

_____

AMAZING ATTRACTION

SCIENCE: Fast-moving air has less pressure than slow-moving air.

STUFF: Two empty soft drink cans, straw, two unopened soft drink cans

What to Do

1. Lay two empty soft drink cans side by side on a level table about 2 finger-widths apart.
2. Hold a straw parallel to the cans about 3 finger-widths above the table, with about a third of the straw between the cans.
3. Blow steadily into the straw.
4. Move the cans farther apart, and repeat Steps 2 and 3.
5. Repeat Steps 1–3 with two full soft drink cans.

What's Going On Here

When you blow between the soft drink cans, they move together. This is opposite of what your intuition tells you should happen. Experience would lead you to assume that the cans would move apart when you blow between them. The reason that the cans move together has to do with air pressure. Before you blow between the cans, the air pressure is the same on both sides of each can. When you blow between the cans, the fast-moving air creates a low air pressure area between the cans. There is higher air pressure on the other side of both cans, so they are pushed together.

Try It!

See how far apart you can place the empty soft drink cans and still make them move together.

See how far apart you can place the full soft drink cans and still make them move together.

With just one empty can on the table, see if you can roll it along the tabletop by continuously blowing on one side of the can.

Try the activity without the straw, using only your mouth to blow between the cans.

AMAZING ATTRACTION

What You Want to Know
What happens when you blow air through a straw between two empty soft drink cans resting on their sides about an inch apart?

What You Think Will Happen
When you blow air through a straw between two empty soft drink cans lying about an inch apart,

       a. the cans will move farther apart.
       b. the cans will move closer together.
       c. the cans will not move at all.
       d. the cans will move slightly upward.

What Happened
What happened when you first blew air between the two empty soft drink cans? How far apart were they?

_____

What happened when you moved the empty cans farther apart and blew air between them? How did you blow differently this time?

_____

What happened when you blew air between two full soft drink cans? How did you blow differently this time?

_____

What It Means
The force of air pressure depends on the speed of the air. Faster-moving air has less pressure. In the drawing below, mark where the faster-moving air is. This is also where the air pressure is less. Explain how the difference in air pressure between the cans and on the outer sides of the cans causes the cans to behave as they do.

_____

_____

_____

BOUNDING BALL

SCIENCE: Fast-moving air has less pressure than slow-moving air.

STUFF: Toilet paper tube, table tennis ball

What to Do

1. Hold a toilet paper tube vertically on one of your hands.
2. Put a table tennis ball into the tube.
3. Place your lips on the edge of the tube, and blow very hard across the top of the tube.

4. Put the ball back into the tube. With your mouth about an inch above the tube, blow very hard directly into it. Close your eyes when you do this because the ball will lightly tap your face.

What's Going On Here

When you blow across the top of the tube, the fast-moving air creates a low-pressure area on the top part of the tube. The air pressure on the bottom part of the tube is higher because the air down there is not moving very fast. The difference in air pressure between the top of the tube and the bottom of the tube causes an upward force on the ball, and it jumps out of the tube. When you blow air directly into the tube, the ball pops out of the tube. Again, there is faster-moving air on the top of the ball than on the bottom of the ball. The air pressure on the bottom of the ball is greater than on the top, and the ball is pushed out of the tube.

Try It!

Try using a paper towel tube.

Try using a heavier ball that fits inside the tube.

Try varying how hard you blow into or over the tube.

BOUNDING BALL

What You Want to Know

What happens when you blow across the top of a toilet paper tube standing on your hand with a table tennis ball in it? What happens when you blow into the same toilet paper tube?

What You Think Will Happen

When you blow air across the top of a toilet paper tube standing on your hand with a table tennis ball in it,

Blow this way.

      a. the ball will stay inside the tube.

      b. the ball will come up a little but not out of the tube.

      c. the ball will pop out of the tube.

When you blow air into a toilet paper tube standing on your hand with a table tennis ball in it,

Blow this way.

      a. the ball will stay inside the tube.

      b. the ball will come up a little but not out of the tube.

      c. the ball will pop out of the tube.

What Happened

What happened to the ball when you blew across the top of the tube?

_____

What happened to the ball when you blew into the tube?

_____

What It Means

There are two forces at work on the table tennis ball in this activity—gravity, which always pulls the ball downward, and air pressure. When you blow into the toilet paper tube, which force is greater?

_____

The force of air pressure depends on the speed of the air. Faster-moving air has less pressure. In the drawing below, mark the place on the ball that has faster-moving air on it (and so less air pressure). Explain how the difference in air pressure on the top and the bottom of the ball causes the ball to behave as it does.

_____

_____

PRESSURIZED PAPER

SCIENCE: Fast-moving air exerts less pressure than slow-moving air.

STUFF: Scissors, ruler, paper, straw

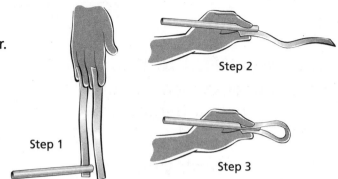

Step 1

Step 2

Step 3

What to Do

1. Cut two strips of paper, each about 1 inch by 8 inches. Hold the two strips between your fingers, one strip between the first and second fingers and the other between the third and fourth fingers. The strips of paper should be vertical, facing each other about an inch apart. Hold a straw horizontally so that one end is between the pieces of paper near the bottom. Gently and steadily blow into the other end of the straw.

2. Place one of the strips of paper so that it overlaps with the end of the straw. The paper should be below the straw. Hold the paper and straw between your thumb and forefinger. Gently and steadily blow into the other end of the straw.

3. Hold one of the strips of paper at the end of the straw as you did in Step 2, but this time place one end of the paper on the bottom of the straw and the other end on the top of the straw. Hold both ends in place with your thumb and forefinger. Blow into the straw.

What's Going On Here

Fast-moving air exerts less pressure than slow-moving air. This concept can be used to explain each of the three activities. All you need to do is determine where the faster-moving air is located. In Step 1 when you blow between the two pieces of paper, the air pressure is less between the papers than it is on the outside of the papers, so they move together. In Step 2, blowing across the top of the paper causes the air pressure to be lower on the top than on the bottom of the paper. This lifts the paper into the air. There is some flapping of the paper due to the constant struggle between gravity and air pressure. In Step 3, the paper moves together. There is faster-moving air between the paper than on the outside of the paper, causing the outsides to push inward. The paper vibrates, producing a sound. The pitch of the sound varies with the thickness and the length of the paper.

Try It!

In Step 1, try blowing at different points between the straws. Try placing the pieces of paper farther apart or closer together.

In Step 2, try differing widths and lengths of paper. Vary how hard you blow over the paper.

In Step 3, try differing widths and lengths of paper, and notice how the pitch of the sound changes.

PRESSURIZED PAPER

What You Want to Know
How does paper react to air being blown over it?

What You Think Will Happen
When you blow air between two narrow pieces of paper hanging close together,

    a. the pieces of paper will move together.
    b. the pieces of paper will move apart.
    c. the pieces of paper will flap together and apart.

When you blow air across the top of a narrow piece of paper,

    a. the paper will move downward.
    b. the paper will move upward.
    c. the paper will flap up and down.

When you blow air between the ends of a narrow piece of paper curved in a U-shape, the upper and lower parts of the paper

    a. will move apart.
    b. will move together.
    c. will flap together and apart repeatedly.

What Happened
What happened when you blew air between two narrow pieces of paper hanging close together?

_____

What happened when you blew air across the top of a narrow piece of paper?

_____

What happened when you blew air between the ends of the narrow piece of paper curved in a U-shape?

_____

What It Means
The force of air pressure depends on the speed of the air. Faster-moving air has less pressure. Pick one of the three steps, and explain how a difference in air pressure caused the paper to move the way it did.

_____

WATER WEIGHT

SCIENCE: Water pressure increases as the depth of the water increases.

STUFF: Hammer, nail, empty soft drink can, masking tape, water, bucket

What to Do

1. Using a hammer and a nail, punch three holes in a soft drink can $\frac{3}{4}$ inch apart, starting about $\frac{3}{4}$ inch from the bottom of the can. The holes should not be in line vertically.

2. Use one piece of tape to cover the holes.
3. Fill the can with water.
4. Hold the can over a bucket.
5. Pull off the tape.

What's Going On Here

When you remove the tape, the water flows from the can because of air pressing down on the top surface of the water. But water is also pushing down with its own pressure. The deeper the water is, the more water there is above it to push down. So water pressure increases with depth; this is similar to air pressure, which also increases as there is more air pushing down from above. You can tell that the water pressure is greater on the bottom than on the top because water from the bottom hole shoots out farthest. It is interesting to see that the path the water takes leaving the can has a curved shape, which is called a parabola. This is the same shape as the path a thrown object takes as it falls to the ground.

Try It!

Try changing the number of holes and their position.

Try using a plastic soft drink bottle.

Try three holes punched horizontally in line about 1 inch apart and $\frac{3}{4}$ inch above the bottom of the can.

Try three holes punched horizontally fairly close together.

Try using a water balloon. Inflate it with water; tie it shut; and punch three holes with a large pin. Make sure you do this outside or over a large bucket.

WATER WEIGHT

What You Want to Know

What happens to water pressure as the depth of the water gets greater? What shape is the path of a stream of water coming out the side of a can?

What You Think Will Happen

In the picture below on the left, draw the paths that you think water from the can will take when tape is pulled off the three holes. In the picture below on the right, mark an X on the line that shows what the path of the water will look like.

What Happened

In the picture below on the left, draw the paths that the water from the can took when the tape was pulled off the three holes. In the picture below on the right, mark an X on the line that shows what the path of the water looked like.

What It Means

Water from a can will squirt out farthest at the level where the water pressure is greatest. What do your observations tell you about where the water pressure is greatest in a can with three holes at different levels?

_____

Explain how the path of water leaving a hole in a can is similiar to the path of a thrown baseball.

CHAPTER 8 | BUOYANCY

Amazing Archimedes Advice

- When an object is placed in water, the force of gravity pushes downward on the object and the buoyant force of the water pushes upward on the object.
- The force of gravity is simply the object's weight.
- The buoyant force is the weight of the water that the object pushes out of the way (displaces).
- Water is at a certain level before an object is placed in it and is at a higher level after an object, no matter how small, is placed in it. The difference between these two water levels can be used to determine the volume of water that was displaced and then, given the density of water, to figure out the weight of the water that is displaced.
- If the weight of the water that is displaced is less than the weight of the object placed in the water, the object will sink. This is part of Archimedes' principle, named after the Greek mathematician and inventor Archimedes (287?–212 B.C.).
- If the weight of the water that is displaced is equal to the weight of the object placed in the water, the object will float. This is the other part of Archimedes' principle.

- Objects that are denser than water will sink if they are not shaped to displace enough water. A chunk of steel will sink to the bottom of a harbor, but shape the chunk of steel into a boat, and it will float because it displaces more water in the new shape.
- Hot water is less dense than cold water, so hot water will float on top of cold water.
- Different liquids have different densities. Denser liquids will sink in less-dense liquids. Saltwater is denser than plain water, so saltwater will sink in plain water.
- An object that floats in plain water will always be able to float in saltwater, and because it will displace a smaller volume of saltwater, it will float higher, with less of its surface below the waterline.

EXTRAORDINARY EGG

SCIENCE: An egg will sink in plain water and float in saltwater.

STUFF: Clear plastic cup (8 or 10 ounces), water, one hard-boiled egg in shell, spoon, salt

What to Do

1. Fill a clear plastic cup about three-fourths full of water.
2. Place a hard-boiled egg in a spoon, and carefully lower it into the water. The egg should sink.
3. Use the spoon to get the egg out of the water.
4. Add salt to the water, and stir, using the spoon, until no more salt will dissolve in the water. This is called a saturated solution.
5. Place the egg in the spoon, and again lower it into the water. The egg should float. You may have to look very carefully because the water will be quite cloudy because of the dissolved salt.

What's Going On Here

When an object is placed in water, the force of gravity pushes down on the object and the buoyant force of the water pushes up on the object. The force of gravity is simply the object's weight, and the buoyant force is the weight of the water that the object pushes aside, or displaces. If the buoyant force is equal to the object's weight, the object will float; if the buoyant force is less than the weight of the object, the object will sink. When the egg is placed in the water, it pushes aside some of the water. If the weight of the water that the egg pushes aside is less than the weight of the egg, the egg sinks. This is the case when the egg is placed in plain water; the egg displaces some water but the weight of the water it displaces is less than the weight of the egg so the egg sinks. When salt is added to the water, the water is denser. A given amount of saltwater weighs more than the same amount of plain water. When the egg is placed in saltwater, it pushes aside almost the same amount of saltwater as it pushes aside of plain water. But since the saltwater weighs more than the plain water, it exerts more of an upward (buoyant) force on the egg and the egg floats.

Try It!

Try the activity with a raw egg.

Try the activity with a peeled hard-boiled egg.

Try floating plain water on top of saltwater. Fill a clear plastic cup halfway with saturated saltwater. Place a spoon inside the cup against the side. Slowly and carefully add plain water to the saltwater by allowing it to flow in along the spoon.

EXTRAORDINARY EGG

What You Want to Know
Will an egg float in plain water? Will an egg float in saltwater?

What You Think Will Happen
When a hard-boiled egg is placed in a cup of plain water, it will
- a. float.
- b. sink.
- c. hover in the middle of the water.
- d. float for a while and then sink.
- e. sink and then float.

When a hard-boiled egg is placed in a cup of saltwater, it will
- a. float.
- b. sink.
- c. hover in the middle of the water.
- d. float for a while and then sink.
- e. sink and then float.

What Happened
What happened when you placed the hard-boiled egg in the cup of plain water?

_____

What happened when you placed the hard-boiled egg in the cup of saltwater?

_____

What did you notice about the appearance of the saltwater?

_____

What It Means
The hard-boiled egg will float if the amount of water that it pushes out of the way as it rests in the water weighs more than the egg. What can you now say about the weight of a certain amount of saltwater compared with the same amount of plain water?

_____

_____

BUOYANT BOAT

SCIENCE: An object will float if it pushes enough water out of the way.

STUFF: Clear plastic cup (8 or 10 ounces), water, paper cup (3 ounces), waterproof marker, 20 pennies

What to Do

1. Fill a clear plastic cup about three-fourths full of water.
2. Float an empty paper cup upright in the water. Mark and label as "1" the level that the water is at, on the outside of the plastic cup.
3. With the paper cup boat floating in the water, add 20 pennies to it, 1 at a time.
4. Mark and label as "2" the new water level, on the outside of the clear cup.
5. Dump all the pennies into the water; mark and label as "3" the new water level.

What's Going On Here

The shape (as well as density) of an object affects whether it will sink or float in water. The shape of an object controls the amount of water that it pushes away, or displaces. If the amount of water that is displaced weighs the same as the object, the object will float. If the displaced water weighs less than the object, the object will sink. When the pennies are in the cup, they displace an amount of water that is equal to their own weight, so they float. As you add more pennies, the cup sinks lower and displaces more water. You can see how much water is displaced by noticing the difference in water level lines that you marked on the plastic cup. When the pennies are out of the paper cup boat and on the bottom of the plastic cup, they displace less water than when they were in the boat, as can be seen by the final water level you marked on the plastic cup.

Try It!

See how many pennies you can put in the paper cup before it sinks.

Make a solution of saltwater, and see how many pennies can go into the paper cup before it sinks.

Place other objects in the paper cup.

Try using a Styrofoam cup instead of a paper cup.

Name _____ Date _____

BUOYANT BOAT

What You Want to Know

How does the water level change when weight is added to a boat? If weight is taken out of a boat and dropped into the water, will the water level be higher or lower than when the weight was in the boat?

What You Think Will Happen

The drawing below shows line 1, the water level in a big cup when a small cup is floating inside with no weight in it. Mark and label line 2 to show where you think the water level will be after 20 pennies have been added to the small cup. Then mark and label line 3 to show where you think the water level will be after the pennies are taken out of the small cup and dropped to the bottom of the water.

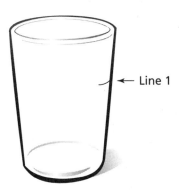

← Line 1

What Happened

Mark and label line 2 to show where the water level was when all 20 pennies were in the cup. Then mark and label line 3 to show where the water level was when all 20 pennies were at the bottom of the water.

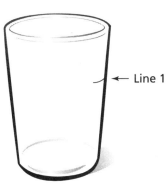

← Line 1

What It Means

What do your observations tell you about how the water level changes when an object floats in a cup and then sinks to the bottom of the water?

_____

FLOATING FOIL

SCIENCE: An object will float if it pushes enough water out of the way.

STUFF: Clear plastic cup (8 or 10 ounces), water, two 6-inch squares of aluminum foil, 20 pennies

What to Do
1. Fill a clear plastic cup about three-fourths full of water.
2. Wrap 20 pennies up in a piece of foil, and drop them into the cup.
3. Remove the package of pennies, and unwrap them.
4. Shape another piece of foil into a boat, and place it in the cup so that it floats without being supported by the sides of the cup.
5. Carefully add the pennies, 1 at a time, to the boat until it sinks.

What's Going On Here
When the pennies are wrapped in foil, they sink to the bottom of the cup of water. The pennies will float in a boat that is designed to push away enough water and to be stable in the water. Two forces are competing here, the force of gravity acting downward on the foil and the pennies and the buoyant force of the water acting upward on them. The pennies and foil sink if the buoyant force is less than the force of gravity (their weight). Archimedes' principle tells us that the buoyant force is equal to the weight of water that is displaced (pushed aside) by the floating object. If the foil boat and pennies push aside enough water to equal their weight, the forces of gravity and buoyancy exactly balance and the boat will float.

 **Try It!**

See how many pennies will float in your boat.

Try other boat designs.

Add about $\frac{1}{4}$ cup of salt to the water, and mix well. See how many pennies will float in your boat.

Try making boats out of other materials.

Try floating other objects in your boat.

FLOATING FOIL

What You Want to Know
How many pennies will float in a boat made from a small piece of aluminum foil?

What You Think Will Happen
How many pennies do you think will float in a boat that you make out of a small piece of aluminum foil?

_____

What Happened
What happened when you wrapped the pennies in a piece of foil and put them in the cup of water?

_____

What ideas did you use to design your aluminum foil boat?

_____

How many pennies went into your aluminum foil boat before it sank?

_____

What It Means
What can you now say about how a boat should be designed to float as much weight as possible?

_____

_____

What changes could you make in your boat design so that the boat would hold more pennies?

_____

_____

What besides the design of a boat affects how many pennies it can hold?

_____

FLIMSY FLOATER

SCIENCE: An object will float in water if the weight of the water it displaces is equal to the weight of the object.

STUFF: Clear plastic cup (8 or 10 ounces), water, waterproof marker, a ball of reusable removable adhesive (used to hold papers on wall; available at school supply stores), fork, 20 pennies

What to Do

1. Fill a clear plastic cup about three-fourths full of water.
2. Mark the water level on the outside of the cup; label the line "1."
3. Drop a ball of reusable removable adhesive into the water. Mark the new water level on the side of the cup; label the line "2."
4. Remove the ball of adhesive from the water, using a fork.
5. Change the shape of the adhesive to make a bowl or a boat that will float in the water. You may have to try a few times to get it to float. When the boat floats, mark the new water level on the outside of the cup; label the line "3."
6. Once you get the boat to float, see how many pennies you can place in the boat before it sinks.

What's Going On Here

According to Archimedes' principle, if the weight of the water that an object pushes aside (displaces) is less than the weight of the object, the object will sink. If the weight of the object is equal to the weight of the water it displaces, then the object will float. When you place the ball of adhesive in the water, it sinks to the bottom. It displaces a tiny bit of water, as you can tell by the line you draw on the cup. But the weight of the water the ball displaces is less than the weight of the ball, so it sinks. When you form the adhesive into a bowl or a boat and place it in the water, you can tell it displaces more water than when it was in a ball because the water level on the cup is higher. The boat floats because the weight of the water it displaces equals the weight of the boat.

 **Try It!**

Try floating the same boat in a solution of saltwater. Will it hold more or fewer pennies? Will it displace more saltwater than plain water?

Try changing the shape of the boat.

FLIMSY FLOATER

What You Want to Know

How does the water level change when a boat floats? How many pennies will float in a boat made of removable reusable adhesive?

What You Think Will Happen

The drawing below shows the water level when a ball of removable reusable adhesive is at the bottom of a cup of water. Make a line on the drawing to show where you think the water level will be when the ball becomes a boat floating in the cup.

How many pennies do you think will float in a boat that you make out of removable reusable adhesive?

_____

What Happened

The drawing below shows the water level when the ball of adhesive was at the bottom of the cup of water. Make a line on the drawing to show the water level when the boat floated in the cup.

How many pennies went in the boat before it sank?

_____

What It Means

What can you now say about how the water level changes when a boat floats?

_____

BOBBING BALLOON

SCIENCE: Cold water is denser than warm water.

STUFF: Five- or 10-gallon fish tank; lukewarm, hot, and cold tap water; bowl; ice; two small balloons

What to Do

1. Fill a fish tank three-fourths full with lukewarm water, and let it stand for about 20 minutes so that the water temperature is near the room temperature.
2. Fill a bowl about one-fourth full of ice. Add cold water to make it half full.
3. Fill a balloon with cold tap water.
4. Place the cold-water-filled balloon into the bowl of ice water for several minutes.

5. Fill another balloon with hot tap water. Fill it to about the same size as the cold-water balloon.
6. Place the cold-water balloon in the fish tank, and observe what happens.
7. Remove the cold-water balloon from the tank. Place the hot-water balloon in the tank, and observe what happens.

What's Going On Here

The cold-water-filled balloon will sink in the room-temperature water. The hot-water-filled balloon will float in the room-temperature water. Very cold water is denser than room-temperature water because the cold-water molecules are packed closer together. Since cold water is denser than room-temperature water, the cold-water-filled balloon sinks in the room-temperature water. The reverse happens with the hot-water-filled balloon. Hot water is less dense than room-temperature water because the molecules are farther apart. Thus the hot-water-filled balloon floats on top of the room-temperature water.

Try It!

Try using water-filled balloons of different sizes for cold and hot water.

Try filling a glass halfway with ice water. Use food coloring to color the water blue. Slowly and carefully add hot water along the sides of the glass to fill it up. The hot water should float on the cold water.

Try filling the fish tank with ice water and placing a balloon with hot water into it. Fill the fish tank with hot water, and place a balloon with ice water into it.

BOBBING BALLOON

What You Want to Know
What happens when you put a balloon filled with cold water into a bowl of room-temperature water? What happens when you put a balloon filled with hot water into room-temperature water?

What You Think Will Happen
When you put a cold-water balloon into room-temperature water, the balloon will
> a. sink to the bottom of the water.
> b. float on top of the water.
> c. first float and then sink.
> d. first sink and then float.

When you put a hot water balloon into room-temperature water, the balloon will
> a. sink to the bottom of the water.
> b. float on top of the water.
> c. first float and then sink.
> d. first sink and then float.

What Happened
What happened when you put the cold-water balloon into room-temperature water?

_____

What happened when you put the hot-water balloon into room-temperature water?

_____

What It Means
Density is a measure of how much matter there is within a given space. A denser liquid will sink in a less-dense one. What can you now say about the density of cold water compared with room-temperature water?

_____

What can you now say about the density of hot water compared with room-temperature water?

_____

© Dale Seymour Publications®

SUNKEN SUBMARINE

SCIENCE: If an object is lighter than water, it will float in water. If it is heavier, it will sink.

STUFF: Six large washers, duct tape, small plastic soft drink bottle, scissors with pointed tip, balloon, clear plastic tubing, rubber band, clay, bucket of water

What to Do

1. Tape six large washers to the outside of a small plastic bottle in a vertical line two washers wide by three washers high.
2. Use scissors to poke four holes spaced evenly vertically in the side of the bottle opposite where you taped the washers.
3. Inflate a balloon to stretch it out. Let the air out, and attach the balloon to one end of a piece of plastic tubing, securing it with a rubber band.
4. Place the balloon and about an inch of the plastic tubing inside the bottle. Use clay around the top of the bottle to hold the tubing in place.

5. Place this submarine in a tub of water large enough to submerge the submarine. Push the submarine into the water so that water comes into the holes you punched in the bottle and the bottle sinks to the bottom. Hold the free end of the plastic tubing so that it isn't submerged.
6. Blow air into the balloon through the free end of the plastic tubing. The submarine should rise to the surface.

What's Going On Here

A submarine rises or sinks according to Archimedes' principle. If the submarine pushes aside (displaces) an amount of water that is equal to its weight, it will float. If it displaces an amount of water that weighs less than the submarine's, it will sink. When the submarine is full of water, it weighs more than the water it displaces, so it sinks.

When you blow air into the submarine, water is pushed out of the submarine. Since air weighs less than water, the air-filled submarine rises to the surface. The air-filled submarine displaces an amount of water that weighs as much as the submarine.

Try operating the submarine without the washers attached.

Try operating the submarine with the washers attached but with the balloon removed from the tubing. When you blow air into the tubing, it will go directly into the bottle.

SUNKEN SUBMARINE

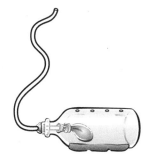

What You Want to Know
What causes a submarine to sink or float?

What You Think Will Happen
A homemade submarine will sink to the bottom of a
tub of water when
        a. air leaves and water enters the submarine.
        b. water leaves and air enters the submarine.
        c. water and air both enter the submarine.
        d. water and air both leave the submarine.

A homemade submarine will rise to the top of a tub of water when
        a. air leaves and water enters the submarine.
        b. water leaves and air enters the submarine.
        c. water and air both enter the submarine.
        d. water and air both leave the submarine.

What Happened
What happened to the air and the water inside when the submarine
sank?

_____

What did you do to change the amount of air inside when you wanted
the submarine to sink?

_____

What happened to the air and the water inside when the submarine
rose?

_____

What did you do to change the amount of air inside when you wanted
the submarine to rise?

_____

What It Means
What do your observations tell you about how the weight of a
submarine can be changed to make it sink or float? When is the
submarine heaviest? When is it lightest?

_____

LITTLE LIFEBOATS

SCIENCE: The carbon dioxide gas in a soft drink causes raisins to rise to the surface.

STUFF: Clear plastic cup (8 or 10 ounces), clear carbonated soft drink, raisins, water, drying cloth, baking soda, measuring spoons, uncooked spaghetti, vinegar

What to Do

1. Fill a clear plastic cup with a clear carbonated soft drink.
2. Drop several raisins in.
3. Observe the motion of the raisins.
4. Clean and dry the cup.
5. Fill the cup halfway with water. Add 1 tablespoon of baking soda, and mix it into the water.
6. Break a piece of spaghetti into bits about $\frac{1}{2}$ inch long.
7. *Slowly* add vinegar to fill the cup. Drop the spaghetti in, and observe what happens.

What's Going On Here

When you drop the raisins into the soft drink, they sink to the bottom because they are heavier than the water. By Archimedes' principle, the amount of water the raisins displace weighs less than the raisins. The soft drink releases carbon dioxide gas in the form of little bubbles. These bubbles collect on the raisins until the raisin-bubble combination is light enough to float to the surface. At this point, the raisin-bubble combination weighs the same as the water it has displaced. Once at the surface, the bubbles start to pop, and carbon dioxide is released into the air. The raisins are again heavier than the water, and they sink. When you use the vinegar and baking soda combination, the spaghetti does the same thing for the same reason. In this case, the vinegar and baking soda react to produce carbon dioxide gas, which acts like little lifeboats.

Try It!

Try watching one raisin and counting how many times it will sink and float.

Instead of raisins, try using balls of clay about the size of rice.

Try using other objects instead of raisins.

Try diet soft drink, and compare it to regular soft drink.

LITTLE LIFEBOATS

What You Want to Know
What happens when you put raisins in a glass of carbonated soft drink?

What You Think Will Happen
When you put raisins in a cup of carbonated soft drink,

       a. they will sink to the bottom and stay there.
       b. they will sink to the bottom, then rise to the top and stay.
       c. they will float on the top.
       d. they will sink and then float repeatedly.

What Happened
What happened when you put raisins in a cup of carbonated soft drink?

_____

What happened when you put spaghetti in a cup of vinegar mixed with baking soda and water?

_____

What did you see happening to the raisin that caused it to sink or float?

_____

How was the action of the spaghetti the same as the action of the raisins?

_____

How was the action of the spaghetti different from the action of the raisins?

_____

What It Means
What can you now say about what causes a raisin to sink or float in carbonated soft drink or spaghetti to sink or float in a mixture of vinegar, baking soda, and water?

_____

What do you think would happen if you put spaghetti in carbonated soft drink or raisins in vinegar, baking soda, and water?

_____

SENSATIONAL SURF

SCIENCE: Less-dense liquids float on denser liquids. Waves are created by some kind of disturbance.

STUFF: Plastic soft drink bottle, water, food coloring, baby oil

What to Do

1. Remove the labels from a plastic soft drink bottle. Fill the bottle about halfway with water. Add a few drops of food coloring to color the water.
2. Fill the bottle to within an inch of the top with baby oil.
3. Tightly screw the lid on the bottle.

4. Lay the bottle on its side, and gently move it from side to side to make waves.
5. Move the bottle up and down slowly, and notice the waves.
6. Move the bottle up and down rapidly, and notice the waves.

What's Going On Here

Baby oil is a less-dense liquid than water and therefore floats on top of water. Moving the bottle up and down causes waves because you are creating a disturbance in the bottle. It is easy to see the crests (high points in the waves) and the troughs (low points in the waves). You can also look at the wavelength, the distance between two crests that are next to each other. When you move the bottle slowly, the wavelength of the waves is relatively large, but when you move the bottle rapidly, the wavelength is shorter. You are putting more energy into the bottle when you move it rapidly, and the resulting shorter-wavelength waves have more energy.

Try It!

Place some glitter or a small object that will float between the two layers of liquid inside the bottle.

Try using other liquids or different temperatures of the same liquid. Hot water will float on cold water. Try it!

Try filling the bottle only a third full of water and about two-thirds full of oil.

Try filling the bottle two-thirds full of water and about one-third full of baby oil.

Put a layer of sand on the bottom of the bottle, and notice what effect the waves have on the sand.

SENSATIONAL SURF

What You Want to Know
How can waves be made in a bottle of baby oil and water?

What You Think Will Happen
When you hold a bottle of baby oil and water sideways,

       a. the baby oil will float on top of the water.

       b. the water will float on top of the baby oil.

       c. the water and baby oil will have totally mixed together.

When you hold a bottle of baby oil and water sideways and move it up and down slowly, the waves will be

       a. about the same as when you move the bottle fast.

       b. closer together than when you move the bottle fast.

       c. farther apart than when you move the bottle fast.

What Happened
When you held the bottle sideways, which liquid was floating on top?

_____

Describe the waves when you moved the bottle up and down slowly.

_____

Describe the waves when you moved the bottle up and down rapidly.

_____

What It Means
What do your observations tell you about how the number of waves in a bottle of liquids changes with how rapidly you move the bottle?

_____

What do your observations tell you about how the distance between the tops (crests) of waves inside a bottle changes with how rapidly you move the bottle?

_____

What can you now say about which is denser—oil or water?

_____

FLOATING FLUIDS

SCIENCE: Liquids of less density float on liquids of greater density.

STUFF: Food coloring (blue and red), four $\frac{1}{4}$-cup measuring cups, water, isopropyl alcohol, clear corn syrup, clear plastic cup (8 or 10 ounces), cooking oil, spoon, four "sinkies" and "floaties" (eggshells, grapes, apple bits, Styrofoam pieces, toothpicks, paper clips, grass, tree twigs, marshmallows, and the like)

What to Do

1. Use food coloring to color $\frac{1}{4}$ cup of water blue and $\frac{1}{4}$ cup of isopropyl alcohol red.
2. Pour $\frac{1}{4}$ cup of corn syrup into a clear plastic cup.
3. Pour $\frac{1}{4}$ cup of water on top of the corn syrup.
4. Pour $\frac{1}{4}$ cup of cooking oil on top of the water. When adding the oil, hold a spoon against the side of the cup, and slowly pour the oil along the side of the cup using the spoon as a guide. This way the liquids won't splash together and mix.
5. Pour $\frac{1}{4}$ cup of isopropyl alcohol on top of the cooking oil. Add it in the same way you added the oil—pouring slowly and using the spoon.
6. Add four sinkies and floaties, one at a time.

What's Going On Here

Many people are familiar with the idea that solid objects can float on water, but this activity stretches the imagination a bit by having liquids floating on liquids! Less-dense liquids will float on denser liquids. Thus alcohol floats on cooking oil, which floats on water, which floats on corn syrup. This is sometimes called a liquid sandwich.

It is interesting to see where the floaties settle in the liquid sandwich. They will sink in the liquids that they are denser than and will float on the liquids they are less dense than. Therefore, an object that has a density between the density of two adjacent liquids will float between the two liquids.

Try It!

Try to find a floatie that will float at each level of liquid.

Make several saltwater solutions with different amounts of salt. Add a different color of food coloring to each sample, and then try to float them on top of each other.

FLOATING FLUIDS

What You Want to Know
Can liquids float on top of each other? Which liquid is lightest? Which one is heaviest?

What You Think Will Happen
Beside each arrow, write *water, corn syrup, alcohol,* or *cooking oil* to show how you think the liquids will settle in the cup.

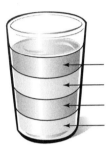

What Happened
Beside each arrow, write *water, corn syrup, alcohol,* or *cooking oil* to show how the liquids settled in the cup. Show where four "sinkies" and "floaties" rested in the stacked liquids.

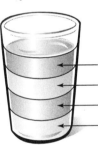

What It Means
Density is a measure of how much matter there is within a given space. A denser liquid will sink in a less-dense one. What do your observations tell you about the densities of water, corn syrup, alcohol, and cooking oil?

_____

_____

What do your observations tell you about the densities of the various objects you tried to float in the stacked liquids?

_____

DROPPER DIVER

SCIENCE: When air is squeezed together, it takes up less space. When water takes the place of air in an object, the object becomes heavier.

STUFF: Plastic soft drink bottle, water, clear plastic cup (8 or 10 ounces), glass dropper (available at pharmacies or at science supply stores)

What to Do

1. Fill a plastic bottle almost completely with water.
2. Fill a plastic cup with water. Fill a glass dropper with a few drops of water, and place it in the cup. Add more water to the dropper until its top floats on the surface of the water in the cup and the glass part of the dropper is underwater. The dropper should be vertical.
3. Put the dropper into the soft drink bottle, and close the lid tightly.
4. Squeeze the sides of the bottle with your hands. The dropper should sink to the bottom.
5. Release your hands. The dropper should rise back to the top. You should be able to make the diver sink and rise repeatedly.
6. If it stops working after a while, try "burping" the bottle by taking the lid off and putting it back on.

What's Going On Here

Initially the dropper is floating at the top of the bottle. The dropper with air and water in it is not heavy enough to sink. When you squeeze the bottle, the dropper sinks. If you observe the dropper very carefully as this is happening, you will notice that the dropper becomes fuller with water. Since the water is heavier than the air it is replacing, the water-filled dropper sinks. When you release the bottle, water leaves the dropper; the dropper becomes lighter; and it rises back to the top. What is actually happening when you squeeze the bottle is that you are decreasing the volume, or amount of room, that the water and air inside it can occupy. The water and air get squeezed into a tighter space. The air, because it is a gas, squeezes together more easily than the water. That is why the air space inside the dropper becomes smaller when you squeeze the bottle. When you release the bottle, the volume increases and the air inside the dropper can expand, pushing the water out and causing the dropper to rise back up to the top.

Try It!

Try filling the bottle three-fourths or half full with water.

Try using a glass jar instead of a plastic soft drink bottle. Cut off the blowing end of a balloon, and stretch the rest of the balloon over the top of the jar like a drum. Push down on the balloon to make the dropper sink.

DROPPER DIVER

What You Want to Know

What happens when you squeeze the sides of a closed, water-filled bottle that has a partially filled dropper inside?

What You Think Will Happen

When you squeeze the sides of a closed, water-filled bottle with a partly filled dropper inside,

       a. water will be pushed out of the dropper, which then sinks.
       b. water will be pushed out of the dropper, which then floats.
       c. water will be pushed into the dropper, which then sinks.
       d. water will be pushed into the dropper, which then floats.

When you let go of the sides of the bottle after you have squeezed it,

       a. water will be pushed out of the dropper, which then sinks.
       b. water will be pushed out of the dropper, which then floats.
       c. water will be pushed into the dropper, which then sinks.
       d. water will be pushed into the dropper, which then floats.

What Happened

What happened to the dropper when you squeezed the sides of the bottle? What happened to the water and air inside the dropper?

_____

What happened to the dropper when you let go of the sides of the bottle after you had squeezed it? What happened to the water and to the air inside the dropper?

_____

What It Means

What can you now say about how you can change whether an object sinks or floats?

_____

What can you now say about how much space the air inside a dropper takes up when it's inside a closed, water-filled bottle and you squeeze the sides of the bottle?

_____

© Dale Seymour Publications®

CHAPTER 9

CENTER OF GRAVITY

Brief Balance Bulletin Board

- The center of gravity of an object is the point about which the mass of the object is evenly distributed.
- An object is balanced when it is supported at its center of gravity.
- You can change the center of gravity of an object by adding mass to or removing mass from some other point on the object.

BALANCED BIRD

SCIENCE: An object will balance at the point about which its weight is evenly distributed.

STUFF: Two pencils (one sharpened, one new and unsharpened), tagboard, scissors, glue, two pennies, clay

What to Do

1. Enlarge the drawing of the bird shown above onto tagboard so that its wingspan is about 5 inches by 4 inches.
2. Cut the bird out, and try to balance it horizontally on the tip of your index finger.
3. Use glue to attach the pennies to the top points of the wings. Try again to balance the bird on the tip of your index finger.

4. Stand the new pencil in a piece of clay so that the eraser end is upward. Balance the bird on the tip of the eraser. The bird's head should be on the eraser tip, and the bird should be almost horizontal.

What's Going On Here

The center of gravity of an object is the point about which the mass of the object is evenly distributed. An object can be balanced when it is supported at its center of gravity. Before you add the pennies to the bird, its center of gravity is in the center of its body. Adding the pennies shifts the center of gravity to the head of the bird, and it can now be balanced at its head.

Try It!

Try to determine the bird's center of gravity with and without the pennies, using the technique described in Steady States on page 226.

Try attaching the pennies to the outside tips of the bird's wings.

Try making other objects out of the tagboard and balancing them by attaching pennies.

BALANCED BIRD

What You Want to Know

Where will a tagboard cutout of a bird balance? Will the point at which the bird balances change when you glue pennies to the top points of its wings?

What You Think Will Happen

On the drawing below, circle the letter for the point where you think a tagboard bird will balance horizontally. Mark an X on the letter for the point where you think it will balance with pennies glued to its wings.

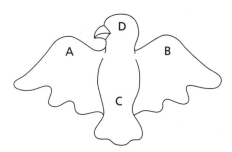

What Happened

On the drawing below, make a mark where the bird balanced when no pennies were on its wings. Label this point "NO" for "no pennies." Then make a mark where the bird balanced when pennies were glued to its wings. Label the point "P" for "pennies."

What It Means

What do your observations tell you about how the balancing point of an object changes when you add weight to the object?

_____

_____

ROCK AND ROLL(ER)

SCIENCE: The distribution of weight in a rolling object can cause some amusing effects.

STUFF: Coffee can with lid, clay, two large rubber bands, wooden board (about 1 foot x 3 feet), several books

What to Do

1. Carefully clean out a coffee can. Mold a piece of clay to lie flat along the seam inside the can. Put the top back on the can.
2. Place a rubber band around the top rim and another around the bottom rim of the can. Be sure to take the twists out of the rubber bands after putting them on the can.
3. Prop a board up on a stack of books to make an incline.

4. Lay the can on the board near the top of the incline, with the clay rotated about 90° away from the board and on the uphill side of the board. Release the can. It should roll uphill and over the edge.
5. Experiment to find the best place to position the can so it rolls over the top edge of the board.

What's Going On Here

Gravity is pulling the clay inside the can straight down. As the clay is being pulled down, the can moves uphill until the clay rests on the board or until the can has gone over the edge. The can will move uphill if it is placed near the bottom or middle of the board also, but it will start wobbling as the clay settles into place on the board. This activity is done with the can at the top of the incline because it seems more mysterious if the can rolls over the edge of the top of the board. The rubber bands on the can give it traction to move on the board.

Try It!

Try the activity without rubber bands on the can.

Try taping a penny to the inside of a small paper cup. Use a book as an incline, and see if the cup will roll uphill.

Roll the can along a level surface, and try to explain its motion.

Try using objects other than clay attached to the seam inside the can.

Try distributing the clay in the can in different ways.

Try other sizes of cans, or paper towel tubes, for the roller.

ROCK AND ROLL(ER)

What You Want to Know
What happens when you place a can on its side so that it can roll along a slanted board?

What You Think Will Happen
When you place a can on its side near the top of a slanted board, it will
>   a. stay there.
>   b. roll down the board.
>   c. roll up the board.

The can will behave the way it does because of
>   a. something having to do with magnets.
>   b. something having to do with weight in the can.
>   c. something having to do with a liquid in the can.
>   d. _____.

What Happened
What happened when you placed the can near the top of the board?

_____

What did you find out caused the can to behave the way it did?

_____

What It Means
What can you now say about how the motion of a rolling object can be changed?

_____

How could you make the can stay still on the board?

_____

How could you make the can roll down the board?

_____

© Dale Seymour Publications®

CLUMSY CLOWN

SCIENCE: An object with a low center of gravity is very stable.

STUFF: Sharp knife, table tennis ball, scissors, ruler, index card (3 inches x 5 inches), tape, clay

What to Do

1. Cut a table tennis ball in half very carefully with a sharp knife.
2. Cut an index card to 5 inches by $1\frac{1}{4}$ inches.
3. Roll the cut index card into a $1\frac{1}{4}$-inch cylinder that will *just* fit onto the cut edge of one of the ball halves. Tape the long edge of the cylinder in place.
4. Tape the bottom edge of the cylinder to the cut edge of the ball half, making sure to keep tape off the bottom of the ball half.

5. Press clay into the ball half attached to the paper cylinder. Use enough to almost fill the ball half.
6. Tape the top edge of the cylinder to the cut edge of the other ball half.
7. Draw a clown face on the cylinder.
8. Try to make the clown lie down. It should stand back upright on its own.
9. Carefully remove the tape from the bottom ball half. Remove the clay, and repeat Step 8.

What's Going On Here

Every object has a center of gravity, the point at which the mass of the object can be considered to be concentrated. In other words, the center of gravity of an object is its balancing point, the point about which the mass of the object is evenly distributed. Clumsy Clown is made with light materials, so without the heavy clay the center of gravity would be in the middle of the cylin-der, midway up from the bottom of the clown. When the clay is added to the bottom of the clown, the center of gravity likewise shifts toward the bottom of the clown. The balancing point is now near the bottom, and therefore the clown balances in an upright position. When you try to lay the clown down, it bobs back upright because it balances well on its bottom.

Try It!

Try rolling the clown down a gentle slope that you can make with a book held at an angle.

Instead of placing clay in the ball half, try placing a large marble inside so that it is free to move around. Then try to roll the clown down a gentle slope made with a book held at an angle.

Try placing the clay in both halves of the ball.

CLUMSY CLOWN

What You Want to Know

How can you change the way that something balances?

What You Think Will Happen

When you place clay in half a table tennis ball on the bottom of a cylinder clown,

       a. the clown will stay lying down when you lay it on its side.

       b. the clown will stand upright when you lay it on its side.

       c. the clown will stand on its head when you lay it on its side.

When you take the clay out of the bottom of a cylinder clown,

       a. the clown will stay lying down when you lay it on its side.

       b. the clown will stand upright when you lay it on its side.

       c. the clown will stand on its head when you lay it on its side.

What Happened

What happened when you tried to lay the clown down when it had clay in the bottom ball half?

_____

What happened when you tried to lay the clown down when it didn't have clay in the bottom ball half?

_____

What It Means

Explain why a cylinder clown balances the way it does when there is clay in the bottom.

_____

Explain why a cylinder clown balances the way it does when no clay is in the bottom.

_____

What could you do to a cylinder clown to make it stand on its head?

_____

What do your observations tell you about how you can change the way something balances?

_____

STEADY STATES

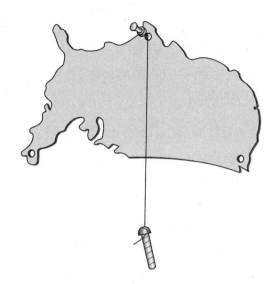

SCIENCE: The geographical center of the United States can be found by using the center of gravity.

STUFF: Glue; tagboard; map of the United States, excluding Alaska and Hawaii (about 8 inches x 10 inches); scissors; two pencils (one sharpened, one new and unsharpened), paper punch; string (24 inches long); push pin; bolt (or other small, heavy object); detailed U.S. map for student page

What to Do

1. Glue a map of the United States, excluding Alaska and Hawaii, onto the tagboard. After the glue is dry, cut out the map.
2. Try to balance the map horizontally on the eraser of a vertically held new pencil.
3. Punch three holes on the edge of the map, far apart and on different sides of the map.
4. Tie a loop in one end of a piece of string, and hang it on the head of a push pin. Tie the other end to a bolt.
5. Hang the map on a bulletin board by putting the push pin through one of the holes you punched in the map. Adjust the push pin so that the map can swing freely on it.
6. Mark where the string hits the edge of the map on the bottom. Remove the map from the board, and draw a line connecting the middle of the punched hole from which you just hung the map and the mark on the map's other edge.
7. Repeat Steps 5 and 6 for the other two holes. The lines should cross close to a common point. Try to balance the map on the eraser of the new pencil at this point.

What's Going On Here

The center of gravity of an object is the point where it balances; all its weight is distributed evenly about that point. The weighted string hangs straight down as gravity pulls on it. The map will hang so that the string runs through the map's center of gravity. By hanging the map from several locations, you can determine its center of gravity, or balancing point.

Try It!

The geographical center of North America is said to be near Rugby, North Dakota. Try finding the center of gravity of North America, and see if it is near Rugby.

Place pieces of clay on the U.S. map to show where major cities are located. Find the population center of gravity by hanging the map from three punched holes.

STEADY STATES

What You Want to Know
Where is the geographical center of the United States (excluding Alaska and Hawaii)?

What You Think Will Happen
The point where a cardboard map of the United States (not counting Alaska and Hawaii) will balance horizontally is called the geographical center of the United States. This point will be located in

- a. Kansas.
- b. Nebraska.
- c. Missouri.
- d. Illinois.

What Happened
What happened when you tried to balance the map on the pencil before you determined its geographical center?

_____

_____

What happened when you tried to balance the map on the pencil after you determined its geographical center?

_____

_____

In which state was the geographical center of the United States (without Alaska and Hawaii) located?

_____

_____

What It Means
The geographical center of the United States (excluding Alaska and Hawaii) is said to be near Lebanon, Smith County, Kansas. Use a detailed map of the United States to determine how far Lebanon, Smith County, Kansas, is from the geographical center that you found in this activity.

_____

How would you determine the geographical center of your state?

_____

What city do you think is nearest the geographical center of your state?

_____

BEAUTIFULLY BALANCED BOX

SCIENCE: An object will balance at the point about which its weight is evenly distributed.

STUFF: Shirt-size cardboard box, marker, bolt (or other small, heavy object), duct tape

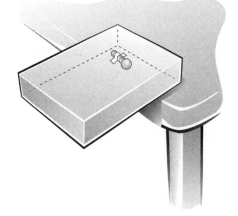

What to Do

1. Place an empty box on a table, and slowly pull it beyond the edge of the table to find the spot where the box no longer balances on the table but falls onto the floor. Keep the short edge of the box parallel to the table edge as you pull it out.
2. Mark a line "1" on the box where it touches the edge of the table when the box *just* balances on the table.

3. Place a bolt in one corner of the box. Hold it in place with duct tape.
4. Place the box on the table so that the bolt of the box is resting on the table. Find and mark as line "2" the point where the box just balances.
5. Repeat Step 4 with the bolt of the box hanging beyond the table. Mark that as line "3."

What's Going On Here

The *center of gravity* of an object is a point about which the mass of the object is evenly distributed. An object can be balanced when it is supported at its center of gravity. Before you add the bolt to the box, its center of gravity is in the center of the box. The point at which it *just* balances on the edge of the table is the middle of the box. Adding the bolt shifts the center of gravity toward the corner of the box where the bolt is located. The point at which the box just balances on the table shifts toward the bolt.

Try It!

Try placing bolts in diagonally opposite corners of the box.

Try making a false bottom for the box so that the activity can be done as a magic trick.

BEAUTIFULLY BALANCED BOX

What You Want to Know
How can you change the way something balances?

What You Think Will Happen
In the drawing below, mark a 1 on the line on which a cardboard box will just balance on the edge of the table. Mark a 2 to show the line on which the box will just balance with a bolt in the end resting on the table. Mark a 3 to show the line on which the box will just balance with a bolt in the end hanging off the edge of the table.

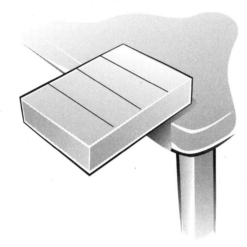

What Happened
Where did the no-bolt box balance?

_____

Where did the box balance when the bolt end was resting on the table?

_____

Where did the box balance when the bolt end was hanging off the edge of the table?

_____

What It Means
What can you now say about how the place where an object balances can be changed?

_____

_____

CHAPTER 10 | CHEMISTRY

Chemical Characteristics

- Matter has both physical and chemical properties. Physical properties include whether it is a liquid, solid, or gas. Chemical properties include how it combines with other matter.
- One chemical reaction is the combination of baking soda and vinegar. In this reaction, baking soda (a solid) combines with vinegar (a liquid) to produce carbon dioxide (a gas).
- Like other gases, the molecules of carbon dioxide move more quickly and take up more space when they are heated.
- Carbon dioxide gas is heavier than air.
- Since fire needs oxygen to burn, carbon dioxide gas can be used to extinguish a flame.
- When materials combine chemically, you often end up with a product that has physical properties totally different from the original materials.
- Polymers are long chains of molecules and are found in a wide variety of natural and synthetic materials. Plastics are examples of polymers.
- Surface tension causes a bubble always to take on a spherical shape, no matter what the shape of the bubble maker.
- Surface tension in liquids can be reduced by adding detergent to the liquid.
- A super water absorber is a chemical that can soak up many times its weight in water.

CORK CANNON

SCIENCE: When vinegar and baking soda mix together, a chemical reaction takes place, producing carbon dioxide.

STUFF: Thumbtack, two pieces of ribbon (each about 8 inches long), cork (or rubber stopper, available at hardware stores), plastic soft drink bottle, scissors, measuring cup and spoons, vinegar, baking soda, piece of paper towel (4 inches x 4 inches)

What to Do

1. Go outside for this activity. It should be done away from other people and buildings.
2. Using a thumbtack, attach a couple of ribbon streamers to the top of a cork.
3. Make sure the cork fits well in the opening of a plastic bottle. Trim a little of it if you need to.
4. Put 1 cup of vinegar in the bottle.
5. Put 1 tablespoon of baking soda in the middle of a piece of paper towel. Roll the towel into a little tube, and twist both ends. This fuel package should fit into the top of the bottle (cannon).

6. Put the fuel package into the cannon and quickly put in the cork. The cork should fit snugly but should not be jammed in.
7. Make sure the top of the cannon is pointed *away* from you. Gently swirl the cannon a bit, and then place it on the ground. Stand back and wait. Be patient. It may take a minute or so.
8. Retrieve the stopper, and quickly put it back in the cannon to see if it can be launched again. See how many times you can launch the stopper by putting it back on after it has popped off.

What's Going On Here

Baking soda and vinegar mix together in a chemical reaction that produces carbon dioxide. The carbon dioxide accumulates inside the bottle (cannon) until the pressure is so great that it pops the cork off the bottle.

Try It!

Try using lemon juice instead of vinegar.

Try using baking powder instead of baking soda.

Try placing the bottle on its side on several straws lying horizontally.

CORK CANNON

What You Want to Know
What happens when you put a stopper in the top of a soft drink bottle that has vinegar and baking soda in it?

What You Think Will Happen
When you put a stopper in a soft drink bottle that has baking soda and vinegar in it,

      a. the stopper will almost immediately pop off.
      b. after a minute or so the stopper will pop off.
      c. after about 5 minutes the stopper will pop off.
      d. after 10 minutes the stopper will still be on the bottle.

When you put the stopper back in right after it has popped off,

      a. the stopper will almost immediately pop off again.
      b. after a minute or so the stopper will pop off again.
      c. after about 5 minutes the stopper will pop off again.
      d. after 10 minutes the stopper will still be on the bottle.

What Happened
What happened when you put the stopper in the cannon that had vinegar and baking soda in it?

_____

What happened when you put the stopper back in the cannon right after it had popped off?

_____

How many times were you able to launch the stopper?

_____

What It Means
What do your observations tell you about what happens when you put a stopper in a cannon with vinegar and baking soda in it?

_____

How could you redesign the cannon so that it would take longer for the stopper to pop off the first time?

_____

BALLOON-BLOWING BOTTLE

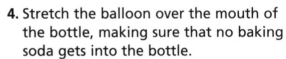

SCIENCE: Vinegar and baking soda mix together to form carbon dioxide.

STUFF: Balloon, piece of paper, measuring cup and spoons, baking soda, small soft drink bottle, vinegar, stopwatch

What to Do

1. Blow a balloon up, and then let the air out.
2. Use a small piece of paper to form a cone-shaped funnel. Using the funnel, pour 1 tablespoon of baking soda into the balloon.
3. Pour 1 cup of vinegar into a soft drink bottle.
4. Stretch the balloon over the mouth of the bottle, making sure that no baking soda gets into the bottle.
5. After you have attached the balloon to the bottle, pull the balloon upward so that the baking soda falls into the vinegar.
6. Time how long it takes the balloon to inflate.

What's Going On Here

When the sodium bicarbonate in baking soda reacts with an acid, like vinegar, carbon dioxide gas is produced. The carbon dioxide gas increases the gas pressure inside the bottle, causing the balloon to inflate. After a short time, the reaction between the vinegar and the baking soda stops, and the balloon does not get any bigger. You can change how large the balloon gets by changing the amount of the materials you use in this activity. By changing the temperature of the materials, you can change how fast the reaction occurs and also how large the balloon inflates.

Try It!

Try using less or more baking soda.

Try using less or more vinegar.

Try the activity using 1 cup of cold vinegar.

Try the activity using 1 cup of hot vinegar.

Try using lemon juice instead of vinegar.

Try using baking powder instead of baking soda.

Try using a larger or a smaller bottle.

BALLOON-BLOWING BOTTLE

What You Want to Know

What happens when you place a balloon over the
top of a bottle that has vinegar and baking soda
in it?

What You Think Will Happen

When baking soda mixes with vinegar in a bottle, a balloon on the top
of the bottle will

       a. immediately get larger.

       b. take about a minute to get larger.

       c. take longer than a minute to get larger.

       d. not change much even after 5 minutes.

What Happened

What happened to the balloon on the top of the bottle when baking
soda mixed with vinegar?

_____

How long did it take for the balloon to get as large as it could?

_____

What It Means

What can you now say about what happens when baking soda mixes
with vinegar in a bottle with a balloon on the top?

_____

What could you change in this activity to make the balloon blow up
larger?

_____

When vinegar and baking soda mix together, a gas called carbon
dioxide is made. Hot gases take up more space than cold gases. Use this
information to describe how you could make the balloon blow up larger
using the same amount of vinegar and baking soda.

_____

_____

FABULOUS FIREFIGHTER

SCIENCE: When vinegar and baking soda mix, they make carbon dioxide gas. The carbon dioxide gas can put out a flame.

STUFF: Measuring cup and spoons, vinegar, plastic soft drink bottle, small candle, matches, baking soda, piece of paper

What to Do

1. Pour 1 cup of vinegar into a soft drink bottle.
2. Light a candle.
3. Measure 1 tablespoon of baking soda. Fold a small piece of paper in half, and hold it at the top of the bottle to direct the baking soda into the bottle.

4. Swirl the bottle around a little, and then lay it on its side with the mouth of the bottle close to but not touching the candle. No liquid should be coming out of the bottle. Be patient. It may take a while, but the candle should go out. If the candle doesn't go out, swirl the bottle around some more and try again.

What's Going On Here

The vinegar and baking soda combine chemically to produce carbon dioxide in the bottle. The carbon dioxide gas pushes the air out of the bottle. That air, however, is not moving fast enough to blow the candle out. The candle goes out when enough carbon dioxide is coming out of the bottle to move over the candle and deprive it of the oxygen from the air that it needs to burn. Carbon dioxide is heavier than air, so as it goes out of the bottle, it sinks toward the table and surrounds the candle flame in the process.

Try It!

Try using baking powder instead of baking soda.

Try using lemon juice instead of vinegar.

Try using a smaller or a larger bottle.

Try using a candle that is taller than the bottle when the bottle is lying on its side.

FABULOUS FIREFIGHTER

What You Want to Know

What happens when you place the mouth of a
bottle with vinegar and baking soda in it near a
burning candle?

What You Think Will Happen

Vinegar and baking soda are mixed together in a soft drink bottle.
When the soft drink bottle is laid sideways with the mouth close to a
burning candle,

       a. the candle will burn more brightly.
       b. the candle will flicker on and off.
       c. the candle will go out.
       d. the candle will not change much at all.

What Happened

What happened when you placed the bottle with vinegar and baking
soda in it near the burning candle?

_____

Was the mouth of the bottle higher or lower than the candle flame?

_____

What It Means

What do your observations tell you about what happens when you
place a bottle with vinegar and baking soda in it near a burning candle?

_____

Vinegar and baking soda mix together to make a gas called carbon
dioxide. Fire needs oxygen to burn. Use these two facts to explain what
happened to the candle flame in this activity.

_____

_____

Carbon dioxide gas is heavier than air. Do you think the activity would
work if you held the mouth of the bottle just below the candle flame?
Explain your answer.

_____

BIGGER, BETTER BUBBLES

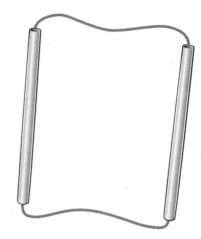

SCIENCE: Surface tension causes bubbles to take shape. Light reflecting off bubbles produces beautiful patterns of color.

STUFF: Measuring cup and spoons, Dawn dishwashing detergent, glycerin (available at pharmacies), water, tub, big spoon, two straws, scissors, string (36 inches long)

What to Do

1. Add $\frac{2}{3}$ cup Dawn dishwashing detergent and 1 tablespoon glycerin to 1 gallon of water in a tub.
2. Gently stir the solution. The solution becomes better with age, so it is a good idea to allow it to sit in an open container for a day before you use it. Store it in a cool place.
3. If the two straws are the flexible kind, cut that part off and use the longer part of the straw.
4. Thread string through both straws, and tie a knot in the end of the string. The easiest way to get the string through the straw is to place a small bit of the string in one end of the straw and carefully suck on the other end of the straw.

5. Put the straws together lengthwise, and dip them into the bubble solution. Pull them out of the solution slowly, still holding them together.
6. Holding each straw in the middle, gently pull them outward to form a rectangle, having the straws be the vertical sides and the taut string be the horizontal sides of the rectangle. Watch the sheet bubble as the bubble solution swirls and finally pops. Notice the beautiful patterns of colored light.
7. Repeat Steps 5 and 6, but this time move the straws smoothly through the air to make a long cylindrical bubble. If the cylindrical bubble lasts long enough, it should change into a spherical bubble.

What's Going On Here

Bubble activities work best on humid days that are not too windy. Surface tension acting on the bubble molecules causes them to form a sphere even though the bubble maker is rectangular. The sheet bubble has a swirling pattern of colored light because of reflection and interference of light.

Light that is reflected off the front surface of the bubble interferes with light that is reflected off the back surface of the bubble to produce the varying patterns of light. These patterns of light are also visible on spherical bubbles.

Try It!

Try using a tennis racket to make bubbles.

Try using your hands to make bubbles by first making a circle using your index fingers and thumbs. Make other shapes with your hands, too.

BIGGER, BETTER BUBBLES

What You Want to Know
What kind of bubbles can you make using two straws and a long piece of string?

What You Think Will Happen
A piece of string threaded through two straws is tied in a knot. This rectangular bubble maker uses the two straws as the short sides and the string as the long sides. When you make bubbles using it,

        a. the bubbles will be long and square.
        b. the bubbles will be long and round.
        c. the bubbles will be short and square.
        d. the bubbles will be short and round.

What Happened
What happened when you placed the straw-and-string bubble maker in the bubble solution and held it up without pulling it through the air?

_____

What happened when you placed the straw-and-string bubble maker in the bubble solution and then pulled it through the air?

_____

What It Means
What can you now say about the kind of bubbles a rectangular bubble maker makes?

_____

How could you redesign the bubble maker to make smaller bubbles? How could you redesign it to make larger bubbles?

_____

_____

How could you design a bubble maker to make many bubbles at one time?

_____

_____

GRISLY GOOP

SCIENCE: When materials combine chemically, you often end up with a product that has physical properties totally different from the original materials.

STUFF: Three paper cups (two 8 ounces and one 3 ounces), warm tap water, measuring cup and spoons, Borax powder (available where laundry products are sold), Elmer's white glue, food coloring (optional), paper towels, sealable plastic sandwich bag

What to Do

1. Fill a large paper cup halfway with warm water. Add 1 teaspoon of powdered Borax at a time, using a craft stick to mix it into the water until no more will dissolve and it just settles on the bottom. Be careful not to get the Borax in your eyes or mouth.
2. Pour $\frac{1}{4}$ cup of Elmer's glue into a second large paper cup. Add 2 tablespoons of water, and stir well using another craft stick. If you like, add a few drops of food coloring.
3. Pour 3 tablespoons of the Borax solution into a small paper cup. Add the Borax solution to the glue and water solution, and stir it rapidly until it gets very thick (about 30 seconds).
4. Take the goop out of the cup; pull it off the craft stick; and knead it with your hands for about 4 minutes. You should do this over a paper towel, keeping the main ball of goop off the towel.
5. Roll the goop into a ball, and try to bounce it.
6. Place the goop on a flat surface, and notice what happens to it after 5 minutes.
7. Hold the goop with both of your hands, and pull it apart very quickly. Put it back together, and this time pull it apart very slowly. Store the goop in a plastic sandwich bag. Be sure to wash your hands thoroughly.

What's Going On Here

Glue is a liquid that has certain familiar properties. Borax is a laundry product used to deodorize and brighten clothes; it comes as a powder that you mix with water to make a liquid. When you mix the liquid Borax with the liquid glue, a chemical reaction occurs, and the resulting product has properties far different from either glue or Borax. What you make is a *polymer*, a material made up of a long chain of molecules. Polymers are very important in our world and are found in a wide variety of plastics.

GRISLY GOOP

What You Want to Know
What happens when you mix Elmer's white glue with
a Borax-and-water solution?

What You Think Will Happen
You are going to mix Elmer's white glue with a
Borax-and-water solution. The resulting goop will be

       a. liquid.   e. powdery.
       b. solid.   f. bouncy.
       c. gaseous.   g. rocky.
       d. sticky.

When you place a ball of goop on a flat surface and don't touch it for
5 minutes, it will

       a. keep its shape.
       b. flatten out.
       c. get bubbles in it.

What Happened
In the table below, list properties of the three materials. Some
properties to consider are color, physical state (solid, liquid, or gas),
texture, and odor. You may think of other properties. *Do not taste the
materials.*

Glue	Borax and water	Goop

What happened when you allowed the ball of goop to sit on a flat
surface for 5 minutes?

_____

What It Means
What do your observations tell you about what happens when you mix
Elmer's white glue with a Borax-and-water solution?

SUPER SOAKER

SCIENCE: A super water absorber can soak up many times its weight in water.

STUFF: Three disposable diapers (super absorbent), two clear plastic cups (8 ounces or 10 ounces), water, bucket, salt, two craft sticks (or tongue depressors), measuring spoons, two cloth diapers, sealable plastic bag (gallon size)

What to Do

1. Open one disposable diaper.
2. Fill a plastic cup with water, and slowly pour the water on the middle of the inside of the diaper. Hold the diaper up by the middle of the edge, and see if any water leaks out. Be sure to work over a bucket.
3. Continue to pour cups of water on the diaper, a little at a time, until it starts to leak when you hold it up. Record the number of cups the diaper held.
4. Repeat Steps 1–3 using a fresh diaper and saltwater. (With a craft stick, stir 2 tablespoons of salt into each cup of water.)
5. Repeat Steps 2–4 using cloth diapers. Record the number of cups the diaper held for each kind of water.
6. Open a third disposable diaper. Pull the inside liner off the diaper. Tear off small

chunks of the cottony material under the liner, and place the chunks in a plastic bag. Get as much of the filler material as you can.
7. Close the bag tightly, and shake it vigorously for about 5 minutes.
8. Carefully pull the filler material out of the bag, shaking each chunk out in the bag as you remove it. When you remove all the filler, you should be left with small particles in the bag that look like sugar. This is the super water absorber that does the work in diapers.
9. Pour the super water absorber into a clean, dry plastic cup. Add $\frac{1}{2}$ teaspoon of water, and stir with a fresh craft stick. Keep adding water until the solution has a gel-like consistency.
10. Sprinkle a little salt on the gel, and stir. Add salt until the gel turns liquid.

What's Going On Here

The super water absorber soaks up many times its weight of water, and it becomes gel-like. When salt is added, the gel changes into a liquid because the salt breaks the bond between the water and the super water absorber. That is why the diaper holds less saltwater than plain water.

SUPER SOAKER

What You Want to Know

How much water can a disposable diaper hold?
What's in a disposable diaper that makes it soak
up so much water?

What You Think Will Happen

When you pour plain water on a disposable diaper, it will soak up _____
cups of water.

When you pour saltwater on a disposable diaper, it will soak up
       a. less saltwater than it did plain water.
       b. more saltwater than it did plain water.
       c. the same amount of saltwater as it did plain water.

When you add a little water to super water absorber (the material that
helps disposable diapers soak up moisture), it will
       a. become a cloudy liquid.
       b. turn into a solid.
       c. turn into a jellylike material.

What Happened

Record your observations in the table.

Kind of water	Number of cups disposable diaper soaked up	Number of cups cloth diaper soaked up
Plain		
Salty		

What happened when you added water to the super water absorber?

_____

What happened when you added salt to the super water absorber?

_____

What It Means

What can you now say about the difference between disposable and
cloth diapers?

_____

BIZARRE BLEND

SCIENCE: A solution made of cornstarch and water has unusual properties.

STUFF: Measuring cup, cornstarch, bowl, water, spoon

What to Do
1. Place $1\frac{1}{2}$ cups of cornstarch in the bowl.
2. Add 1 cup of water, and stir the solution.
3. Slowly press your index finger into the solution.
4. Try poking your index finger very rapidly into the solution.
5. Slowly push the bottom side of the eating end of the spoon to the bottom of the solution; then try to quickly slap the flat end of the spoon into the solution.
6. Try mixing the solution with the spoon quickly and then slowly.
7. Pick up a chunk of the solution, and quickly try to pack it into a ball. Then let the ball rest in your open hand.

What's Going On Here
The cornstarch-and-water solution has unusual properties. As you increase the pressure on the solution, it becomes stiffer and does not flow very easily. This is what happens when you try to quickly push your finger or the spoon into the solution and when you try to quickly form it into a ball.

The solution becomes less stiff and flows more easily when you decrease the pressure on it. This is what happens when you slowly lower your finger or the spoon into the solution or let a chunk of it rest in your hand.

Try It!

Try adding different amounts of water to the cornstarch to get different properties of the solution.

Try using other liquids instead of water.

BIZARRE BLEND

What You Want to Know
What happens when you mix cornstarch with water?

What You Think Will Happen
When you slowly press your index finger into a solution of cornstarch and water, your finger
　　　a. will slowly go to the bottom of the bowl.
　　　b. won't be able to go to the bottom of the bowl.

When you poke your index finger very rapidly into a solution of cornstarch and water, your finger
　　　a. will rapidly go to the bottom of the bowl.
　　　b. won't be able to go to the bottom of the bowl.

What Happened
In the table below, list properties of the three materials. Some properties to consider are color, physical state (solid, liquid, or gas), texture, and odor. You may think of other properties.

Cornstarch	Water	Bizarre blend

What happened when you slowly pressed a finger into the solution?

_____

What happened when you rapidly poked a finger into the solution?

_____

What happened when a ball of bizarre blend rested in your open hand?

_____

What It Means
What can you now say about the properties of bizarre blend?

_____

MARVELOUS MOVING MILK

SCIENCE: Surface tension in milk can be lessened by adding detergent.

STUFF: Whole milk, pie pan, Sunlight dishwashing detergent, paper cup (3 ounces), food coloring (four colors), dropper

What to Do

1. Let whole milk stand unrefrigerated for about an hour, until it reaches room temperature.
2. Pour milk into a pie pan to a depth of about $\frac{1}{2}$ inch. Pour about $\frac{1}{2}$ inch of detergent into a small cup.
3. Drop four drops of food coloring on the surface of the milk. Use different colors, and try to place the drops so that they are not touching each other.

4. Add a few drops of detergent at other locations on the milk surface, using a dropper.
5. As the reaction slows down, add a drop of detergent to a place on the surface of the milk where the food coloring appears to be concentrated.

What's Going On Here

A number of physical science principles are involved in this activity. The detergent breaks the surface tension of the fat particles in the milk. The detergent also sinks to the bottom of the pie pan because of its greater density. The food coloring dissolves and mixes with the milk as the detergent is both sinking to the bottom of the pie pan and breaking up the fat particles in the milk. As the food coloring mixes with the milk, its motion produces the beautiful patterns of swirling color that you observe.

Try It!

Try using a clear plastic cup instead of a pie pan.

Try using other liquids, such as skim milk, cream, water, or juice.

Try using a different temperature of milk, hot or cold.

Try using different kinds of detergent, soap, or shampoo.

MARVELOUS MOVING MILK

What You Want to Know
What happens when you place drops of food coloring and then drops of detergent in a pan of milk?

What You Think Will Happen
When you put a few drops of detergent into a pan of milk with a few drops of unmixed food coloring already in it,

       a. the milk and food coloring will swirl around.

       b. the drops of food coloring will sink to the bottom of the pan.

       c. the detergent will react with the milk so that the food coloring can not mix in.

       d. the detergent will sink to the bottom of the pan, and the drops of food coloring will stay in small circles on the top.

What Happened
What happened when you put a few drops of detergent into the pan that had milk with a few drops of unmixed food coloring in it?

_____

_____

After the reaction slowed down, what happened when you added another drop of detergent?

_____

_____

What It Means
What can you now say about what happens when you drop detergent into milk?

_____

_____

What kind of milk, food coloring, and detergent could you use to test whether other kinds of *detergent* work in this activity?

_____

_____

What kind of milk, food coloring, and detergent could you use to test whether other kinds of *milk* work in this activity?

_____

_____

GLOSSARY

absorb To take in or soak up.

air pressure The weight of the air pushing down on us; the force of air molecules bouncing off something.

air resistance A force that opposes the motion of objects through air. As an object moves through air, it is constantly bumping into air molecules that resist the object's motion.

angle The opening between two lines that meet at a point.

angle of incidence The angle that light strikes a surface.

angle of reflection The angle that light bounces off a surface.

angular momentum The momentum of an object moving in a circular path. It is the product of the object's mass, speed, and radius of motion.

Archimedes' principle An object will float if the weight of the water it displaces is equal to the weight of the object; the object will sink if the weight of the displaced water is less than the weight of the object. The principle is named after its discoverer, Greek mathematician and inventor Archimedes (287?–212 B.C.).

attract To pull close to. Unlike electrical charges attract each other.

Bernoulli's principle Part of the principle states that faster-moving air exerts less pressure than slower-moving air. The principle is named after its discoverer, Swiss mathematician David Bernoulli (1700–1782).

buoyant force The force of water pushing up on an object. It is equal to the weight of the amount of water that is displaced by the object.

carbon dioxide A molecule made up of one carbon atom and two oxygen atoms. Carbon dioxide is a gas at room temperature.

center of gravity The point about which the mass of an object is evenly distributed.

centripetal force The force on an object that is moving in a circular path. The direction of this force is toward the center of the circle.

charging, electrical The act of rubbing electrons off one object and onto another, as when you rub a balloon on your hair.

concave mirror A mirror that is curved inward so that it looks like a cave.

conductor, electrical A material through which electrons easily flow.

converging lens A lens that is thicker in the middle than at the edges.

convex mirror A mirror that is curved outward.

compressed Pushed tightly together.

crest The high point of a wave.

cylindrical lens A lens that is shaped like part of a cylinder. It focuses light in one direction only.

density How much matter there is within a given space; mass per volume. A box filled with feathers is less dense than the same box filled with books.

diaphragm A large muscle located below the lungs. Its movement causes air to be taken into our lungs and pushed out of our lungs.

diffraction The spreading out of waves as they go around an obstacle or through an opening.

dipole An object with two magnetic poles. Atoms in a magnet are tiny magnetic dipoles.

disperse To break up. A prism disperses white light into its component colors of red, orange, yellow, green, blue, indigo, and violet.

displace To push aside.

drag The force that opposes the forward motion of an airplane. Drag is caused by air resistance.

echo The reflection of sound waves off an object.

electric current The flow of electrons through a material.

electromagnet A magnet that is made by wrapping a wire around a piece of iron and then running an electric current through the wire.

electron A negatively charged particle that occupies space around the central part (nucleus) of an atom.

energy The ability to do work.

focal length The distance from a lens or a mirror to the point where light from a distant source is brought to a focus.

focal point (or **focus**) The location where light rays from a distant object come to a point after passing through a lens or being reflected by a mirror.

force A push or a pull.

frequency The number of times in 1 second that a vibrating object moves back and forth; the number of waves that pass a point in 1 second.

friction A force caused by irregularities in the surfaces of objects that are moving over each other. The direction of the force of friction is always opposite to the motion of an object.

fuse A device that breaks (opens) a circuit when too much current flows through the circuit.

gravity The attraction between two objects due to their masses. The force of gravity that attracts a person toward the earth is the person's weight.

gravity, law of Every object in the universe attracts every other object in the universe with a force. This force is greater for greater masses, and it decreases as the distance between the objects increases.

heat The energy that flows from one object to another because of a difference in temperature; the energy due to the motion of molecules. When an object is heated, the molecules move faster. When an object is cooled, the molecules move slower.

horizontal symmetry The property that an object appears the same when it is flipped over a horizontal line.

image distance The distance from a lens or a mirror to an image of an object formed by the lens or mirror.

inertia The measure of an object's resistance to a change in its motion, in either direction or speed.

insulator, electrical A material through which electrons do not easily flow.

intensity Brightness of light; loudness of sound.

interference The encountering of two waves. If they are the same wavelength and meet with the crest of one matching the trough of the other, they cancel each other out. If the crests match each other, they reinforce, making a wave that is bigger than either of the first two.

kinetic energy The energy of motion. The kinetic energy of an object increases as its mass or its speed increases.

lift The force on an airplane's wings that is due to the difference in air pressure on the top and on the bottom of the wings.

light shield An object or objects that are designed to block out unwanted light.

magnetic field The area around a magnet. Objects attracted to the magnet will be pulled toward the magnet more strongly where the magnetic field is stronger.

magnification The ratio between the image size and the object size. The larger the image is than the object, the greater the magnification.

mass A measure of the amount of matter in an object.

matter Anything that has mass and occupies space.

momentum The short form for *linear momentum*, the product of an object's mass and speed.

negative (electrical charge) The condition in which an object has an excess number of electrons.

neutral (electrical charge) The condition in which an object has the same number of positive and negative charges, so it appears as if the object has no net charge.

Newton's first law of motion An object stays at rest or in motion at a constant speed in a straight line unless it is acted on by a force. The law is named after its discoverer, English astronomer, scientist, and mathematician Sir Isaac Newton (1642–1727).

Newton's second law of motion The more you push or pull on an object, the greater effect you will have on changing its speed or direction of motion. The object will move in the direction that you pushed or pulled it. The law is named after its discoverer, English astronomer, scientist, and mathematician Sir Isaac Newton (1642–1727).

Newton's third law of motion For every force on one object, there is an equal (in size) and opposite (in direction) force on another object. The law is named after its discoverer, English astronomer, scientist, and mathematician Sir Isaac Newton (1642–1727).

north pole The part of a freely moving magnet that points toward the earth's geographical north pole. The north pole of one magnet repels another magnet's north pole and attracts its south pole.

object distance The distance from a lens or a mirror to an object.

opaque Absorbing all the light that strikes. No light passes through an opaque object.

parabola The curved path that a thrown object takes as it falls to the ground.

parallel circuit A circuit in which the components are connected in such a way that any component makes its own complete circuit.

pendulum An object that swings back and forth. An example is a washer attached to a string.

pitch The highness or lowness in the tone of a sound. The higher the frequency, the higher the pitch.

plano-convex lens A lens that is curved outward on one side and is flat on the other side. It is one kind of converging lens.

polymer A long chain of molecules made by connecting a large number of molecules together.

positive (electrical charge) The condition in which an object has a deficiency of electrons.

potential energy Stored energy that may be changed to kinetic energy. One kind of potential energy is the result of an object's position—for example, its height above the ground. If the object falls to the ground, the potential energy changes to kinetic energy as it falls.

pressure The ratio of force per area that the force is acting on.

reflection The bouncing back of light from a surface.

reflection, law of The angle that light strikes a surface is the same as the angle that it is reflected from the surface. The angle of incidence equals the angle of reflection.

refraction The bending of light as it passes from one material to another, as from glass to air or glass to water.

repel To push away. Like electrical charges repel each other.

resistance, electrical The opposition to the flow of an electric current through a material. Thin wires have more electrical resistance than thicker wires of the same length.

revolution The motion of an object that is moving in a circle around another object.

right-hand rule A principle that gives the direction of the magnetic field around a current-carrying wire. Pretend that you are grasping the wire with your right hand and pointing your thumb in the direction the current moves as it goes from the positive terminal to the negative terminal of the battery. Your fingers will be pointing in the direction of the magnetic field.

scatter To send off in many different directions.

series circuit A circuit in which the components are connected to one another in such a way that the same electric current flows through all the components.

south pole The part of a freely moving magnet that points toward the earth's geographical south pole. The south pole of one magnet repels another magnet's south pole and attracts its north pole.

spectrum All the colors that are in white light: red, orange, yellow, green, blue, indigo, and violet.

speed A measure of how fast an object is moving. Average speed is determined by dividing the distance an object travels by the elapsed time.

static electricity The buildup of electrical charge, either positive or negative, on an object.

super water absorber A chemical that can absorb many times its weight in water.

surface tension The attraction of the molecules in some materials to one another. Surface tension in water allows you to fill a glass slightly over the top.

temporary magnet A material that behaves like a magnet for a short time.

thrust The forward force on an airplane provided by the engines.

translucent Letting some light pass through but not enough to allow you to clearly see objects beyond. Wax paper is an example of a translucent material.

transmission The passing of light through an object. For example, glass and clear plastics transmit light well.

transparent Letting enough light pass through to allow you to see objects beyond. Transparent objects transmit almost all the light through them.

troughs The low points in waves.

vacuum Space that has no matter in it.

vertical symmetry The property that an object appears the same when it is flipped across a vertical line.

vibration Moving back and forth.

volume The amount of space that an object occupies; the loudness or softness of sound.

water pressure The weight of water pushing down on an object.

wavelength The distance between two adjacent peaks in a wave.

weight The force of gravity acting on an object.

work What is done when a force is applied to an object and the object moves in the direction of the force.

SELECTED BIBLIOGRAPHY

Allen, Maureen, et al. *Electrical Connections.* Fresno, Calif.: AIMS Education Foundation, 1991.

Allison, Linda, and David Katz. *Gee Whiz: How to Mix Art and Science, or the Art of Thinking Scientifically.* Boston: Little, Brown, 1983.

Bentley, Joan, and Linda Hobbs. *How to Do Science Experiments with Children.* Monterey, Calif.: Evan-Moor, 1994.

Herbert, Don. *Mr. Wizard's Supermarket Science.* New York: Random House, 1980.

Hoover, Evalyn, et al. *Mostly Magnets.* Fresno, Calif.: AIMS Education Foundation, 1991.

Hoover, Evalyn, et al. *Principally Physics.* Fresno, Calif.: AIMS Education Foundation, 1991.

Levenson, Elaine. *Teaching Children about Science: Ideas and Activities Every Teacher and Parent Can Use.* Englewood Cliffs: Prentice-Hall, 1985.

Liem, Tikl. *Invitations to Science Inquiry.* Chino Hills, Calif.: Science Inquiry Enterprises, 1987.

Murphy, Pat, and Suzanne Shimek. *Exploratorium Science Snackbook.* San Francisco: Exploratorium Teacher Institute, 1991.

Smith, Robert. *Hands-On Science: The Science Series.* Grand Rapids, Mich: Instructional Fair, 1989.

VanCleave, Janice. *Physics for Every Kid: 101 Easy Experiments in Motion, Heat, Light, Machines, and Sound.* New York: John Wiley, 1991.

VanCleave, Janice. *200 Gooey, Slippery, Slimy, Weird, and Fun Experiments.* New York: John Wiley, 1993.

VanCleave, Janice. *201 Awesome, Magical, Bizarre, and Incredible Experiments.* New York: John Wiley, 1994.

Walpole, Brenda. *175 More Science Experiments to Amuse and Amaze Your Friends.* New York: Random House, 1990.

Walpole, Brenda. *175 Science Experiments to Amuse and Amaze Your Friends.* New York: Random House, 1988.

INDEX